P9-CLZ-944
DISCARD

DAYHIKER

Other Books by Robert S. Wood

Desolation Wilderness
**Pleasure Packing*
**Mountain Cabin*
Good-Bye, Loneliness!
**The 2 Oz. Backpacker*
**Whitewater Boatman*
Homeopathy, Medicine That Works!

*Available from Ten Speed Press

DAYHIKER

WALKING FOR FITNESS, FUN AND ADVENTURE

Robert S. Wood
illustrations by Warren Lloyd Dayton

Ten Speed Press
Berkeley, California

Ten Speed Press
Box 7123
Berkeley, CA 94707

Illustrations by Warren Lloyd Dayton
Text and cover design by Fifth Street Design, Berkeley, California.

Library of Congress Cataloging-in-Publication Data

Wood, Robert S. (Robert Snyder), 1930–
Dayhiker / Robert S. Wood.
drawings by Warren Lloyd Dayton
p. cm.
Includes bibliographical references
ISBN 0-89815-406-5
1. Hiking. 2. Hiking—Equipment and supplies. I. Title.
GV199.5W66 1991
688.7'651—dc20 90-28282
CIP

Printed in the United States of America

1 2 3 4 5 — 95 94 93 92 91

"Never lose your love for walking," wrote Danish philosopher Soren Kierkegaard. "I've walked myself into my best ideas. I've walked myself out of bad ones. I've walked myself into health and avoided disease."

You can, too!

I dedicate this book to my companions of the trail, who have added so much to my enjoyment of the wilds.

Table of Contents

INTRODUCTION
Dayhiking

Welcome to the widening world of dayhiking! In the past few years, Americans by the millions have discovered its allures. And more are joining the ranks every day. Backpackers and joggers are making the switch in droves. Dayhiking is booming because it's so easy and rewarding. You don't have to run or carry a backpack. You just put on your shoes and go walking in the wilds.

Dayhiking offers something for everyone: stimulation, relaxation, radiant good health, mental enrichment, freedom from heavy packs, solitude, companionship, weight loss, accomplishment, escape, communion with nature, fitness, peace, personal growth. And more.

This little book contains all you need to know to safely dayhike in comfort, whatever your particular goals. You'll find time-tested tips and techniques for sailing confidently and securely through the wilds, staying out of trouble, and squeezing the maximum pleasure from your walks. It will maximize your comfort and minimize anxiety, while helping you deal with outdoor adversity.

The book is organized for easy access to specific material. You can find what you need without reading the whole book. Each of the ten chapters is divided into topics important to the dayhiker. You'll find them listed in three different places:

(1) the Table of Contents, (2) the chapter headings, and (3) as subheadings within each chapter.

Besides helping you gain all of dayhiking's myriad benefits, I'm pleased to introduce you to two great boons to a walker's health and happiness: an eating style that will double your energy and fitness on the trail while cutting your weight—without leaving you hungry (chapter 6), and an introduction to homeopathy, the most effective system of medicine known to mankind (chapter 8). Homeopathy is as good at curing chronic afflictions as it is dependable for first aid on the trail.

In addition, you'll learn the secrets of living and walking in the wilderness for days or weeks on end without the need to carry a backpack (chapter 7). And by mastering the fundamentals of navigation (chapter 10) you'll always know where you are and feel at home in the woods. The dayhiker's best friend, low-bulk, little-known vapor barrier warmth, is discussed in chapter 3. The colder it is, the better it works. Plus the collected insider tips and techniques that will ensure your safety and comfort in the wilds (chapter 9).

Please forgive me, ladies, for use of the masculine gender throughout. I'm delighted that more women than ever—singly and in groups—are out walking the wilds.

I hope this book helps you enjoy easier, safer, and more comfortable journeys wherever you dayhike. Let's go walking together through the pages that follow.

Pollock Pines, California
March 1991

ONE
Why Dayhike?

My Definition . . . The Ideal Form of Exercise . . . Dayhiking or Backpacking? . . . Further Benefits

My Definition

Dayhiking embraces every kind of walking from street hiking to peak bagging, from an exuberant five-minute jaunt on your lunch break to three weeks of daily hut-to-hut hikes in the Alps. In between, it includes aerobic exercise for maximum fitness and weight loss, hour-long rambles in the park, strolls down country lanes, serious half- or full-day treks in the wilds, trail walks, cross-country scrambles, even mountain climbing. It's any kind of walking except backpacking, because dayhikers, by definition, are free of overnight packs.

Cross-country skiing qualifies because your legs propel you through the outdoors. I even include river rafting because it gets you to the wilds without an overnight pack and permits you to walk unencumbered. Dayhiking offers surprising rewards for the city-bound, and in its highest form it's unbeatable for fully and freely experiencing the wilds unburdened. Develop an addiction to dayhiking, and your fitness for life is practically assured. So is any needed weight loss. Regular walkers aren't fat.

Perhaps just as valuable are the mental and emotional benefits of tranquility, stress relief, heightened awareness of the world's natural beauty, the chance for uninterrupted meditation (or music or learning if you wear a Walkman), escape from your cares and worries, and stimulation of your problem-solving abilities. Dayhikers enjoy it all.

The Ideal Form of Exercise

And it's so simple and easy, relaxing and enjoyable, safe and beneficial, available and companionable. That's why America is enjoying a dayhiking boom as joggers and backpackers make the switch. Walking is the ideal form of exercise. Runners worry about shinsplints and muscle tears. Cyclists have to dodge cars. Swimmers need a pool. Tennis players need a court and a partner. But dayhikers simply set forth. The whole outdoors is their arena. And when it's snowy or cold, they simply move indoors to heated malls, corridors, and gyms.

"More people are walking than ever before," stated *Mother Earth* magazine recently, "but they are dayhiking, not backpacking." As evidence, we now find daypacks, beltpacks, and fannypacks in our supermarkets and a wild proliferation of "walking" shoes in shoe stores.

At first glance, dayhiking seems nothing more than a stroll that ends when the sun goes down and the walker returns to civilization. It's true that a dayhiker must find shelter for the night, since he didn't carry it with him on his back. But that doesn't mean the fun has to stop. Far from it. Return needn't be an anticlimactic end to an outdoor adventure. Instead it can be a celebration of the day's adventures mixed with delicious anticipation of the day to come.

Dayhiking or Backpacking?

I've been a backpacker for forty years, and I've written two successful books on the subject. Don't imagine that I've given it up. I still happily backpack when I want to see country I can't reach any other way, but I much prefer dayhiking because it's so free and easy. There isn't any conflict between dayhiking and backpacking. They complement one another and are easily combined for maximum satisfaction in the wilds. Although they're compatible, backpacking is a lot more work and requires considerable advance planning and preparation, which helps explain the swing toward dayhiking. Dayhikers don't have to keep their eyes riveted to the trail. They're free to savor the country they've come to see.

As I've grown older, I hope I've grown smarter. I've found ways to avoid the excessive labor and preparation of backpacking, yet see the same country—*really* see it. The next time you contemplate a weekend excursion, ask yourself whether backpacking will really maximize the experience. Or should you consider dayhiking instead?

Backpackers, of course, are often dayhikers, too, and vice versa, but contrasting the two approaches to seeing specific places can produce some startling revelations and sharpen your perspectives. Dayhikers tend to cover more miles with less effort, see more country, climb more peaks, enjoy more cross-country travel, hike more often and more spontaneously, take more pictures, see more wildlife, and catch more fish. They just plain have more fun because they don't have to carry a house on their backs.

The average serious backpacker with two weeks of annual vacation probably spends fewer than ten nights a year camping on two to three trips. By contrast, a serious dayhiker with the same job might typically find time to take fifty hikes—that's only one a week. Is sleeping on the ground really worth the extra effort? I'm not knocking backpacking where it's the best or only way to go. I'm just trying to make you question your motives and focus on your goals, to help you honestly consider how to squeeze the most from your weekend in the wilds. So the next time you think about backpacking to Moosejaw Lake, ask yourself if dayhiking might not serve you better.

Dayhiking lets me have my cake and eat it, too, allowing me to overcome its lone drawback: going home at night. Strategies for sidestepping that problem have opened new vistas, new country. They've brought me more opportunities for enjoying the wilderness with less effort—without a pack—than I ever dreamed existed. I've written this book to share some of those discoveries.

Further Benefits

Dayhiking frees me from the trails, permitting me to go cross-country and see the true wilderness that doesn't begin until you leave the man-made path behind. It permits me to see more country on any given day than can be seen if I'm bent beneath an overnight pack. I go farther and faster and higher, covering more ground with less effort, often climbing peaks on mere whim because they're so close. I'm freer and fresher and better able to savor the solitude I seek because I travel so light.

Let me be more specific. I'm free of the need to find an appealing, suitable, legal, safe place to camp . . . not always easy these crowded days. I don't have to worry about the safety of my water supply, or the gear I leave in camp. I'm largely independent of the weather. I can—and often do—go on satisfying dayhikes in storms that would be miserable to camp in. Backpacking takes meticulous planning, while dayhiking can be spontaneous and free.

That's why nowadays—like millions of outdoor Americans—I do most of my walking on dayhikes.

TWO

Five Kinds of Dayhikes

Aerobic Dayhikes . . . Compute Your Training Level . . . Mental & Emotional Benefits . . . Quickies . . . Time to Kill? Go for a Walk! . . . Half-Day Hikes . . . Beware Road Walking . . . All-Day Hikes . . . Weekend Walks . . . Combo Trips

For convenience, I've divided dayhiking into seven different categories: aerobics, quickies, half days, full days, combinations, extended trips, and climbs. We'll consider the attributes and virtues of each in that order. The first five appear in this chapter; extended dayhikes and climbs are in chapter 7.

During the recent fitness boom, it's been well documented by experts that for health, happiness, weight loss, and long life, the human organism requires frequent aerobic exercise for at least twenty continuous minutes. A workout every other day is ideal, providing the body with the optimum balance of rest and exercise. Aerobic exercise works the heart, stirs the blood, stretches the lungs, gently massages the organs, stimulates elimination of toxic material, burns excess fat, strengthens the muscles, and brings a healthy glow to the skin.

Shelffuls of books have been written on how to exercise aerobically and how to make yourself do it. They all agree that aerobic exercise is the most efficient and effective way to spend twenty minutes if you want to maximize your fitness and weight loss. The problem is sticking with it. The most comprehensive list of strategies I know for keeping at it can be found in chapter 4 of Mort Malkin's *Walking—The Pleasure of Exercise.* "Malkin's Magic Motivators" will give you the strength to go on if you're inclined to procrastinate. They made me want to put down the book and go walking.

Aerobic walking is almost too good to be true. It takes no training, no equipment, no partner, no sports facility, very little time, and not much effort. Almost anyone can do it, even many of the aged, ailing, and injured. In fact, walking is increasingly prescribed by physicians as the basis of rehabilitation programs of all kinds. Nearly everyone benefits from walking. There's practically no down side, although an aerobics program imposes strict requirements on the walker.

While everyone knows how to walk, a few pointers may improve your form and power. Serious walkers lengthen their stride—and therefore their speed and efficiency—by reaching forward slightly with the hip at each step. By thrusting forward with the hips you use the powerful large muscles of the buttocks, hips, and thighs, rather than relying on the smaller, more easily tired calf muscles. Walk erect, holding your head high. Swing your arms vigorously in rhythm with your legs, but don't bounce. Keep your feet and your legs pointed straight ahead. And smile. When you're doing it right it feels powerful and good.

Aerobic Dayhikes

There are various formulas for measuring the level of your exertion, and there are many fancy, expensive programs and machines to help you do aerobic exercises. Finding the needed motivation to exercise is a major problem for habitually sedentary people surrounded with tempting alternatives. It's so much easier for most of us to collapse in front of the TV set and eat junk food. For most dayhikers, however, motivation isn't an overwhelming problem. We have somehow acquired a yen for fresh air, sunshine, challenging trails, beautiful country, and the good feelings (mental and physical) that come from a vigorous stroll in the country.

By definition, to qualify as "aerobic," exercise must raise your heart rate to your personal "training heart rate" (see below) and keep it there continuously for a minimum of twenty minutes (plan on thirty to be safe). That's it. There are various forms of exercise that qualify, but there's a consensus among experts that vigorous walking is the easiest way to achieve and maintain an aerobic level of exercise for most people. The "continuous" part is essential. Drop below the prescribed heartbeat level and you've blown it. Then you have to start over to gain the desired effect. That's why you see people jogging in place on curbs, waiting impatiently for traffic lights to change. They don't want to risk dropping below their aerobic training level and starting over.

Compute Your Training Level

Here's an easy way to compute the needed heartbeat rate for your age. You need a wristwatch that counts seconds, and you need to be able to take your pulse, that is, count the number of times your heart beats during a period of ten seconds. The easiest places to find your pulse are in the insides of your wrists and to either side of your Adam's apple in your neck.

YOUR TRAINING ZONE

AGE	HEARTBEATS/TEN SECONDS
10–19	25–30
20–29	24–28
30–39	22–27
40–49	21–25
50–59	20–24
60–69	19–22
70–79	18–21
80–89	16–19

Here's how to use the chart. Opposite your age you'll find the number of heartbeats per ten-second counting period that you need to maintain for twenty continuous minutes to qualify for a genuine aerobic workout. The lower number is the minimum. You can safely go a little higher than the maximum without losing any benefits, but you won't gain any, either. For instance, if you're thirty-seven you need to feel at least twenty-two heartbeats in a span of ten seconds when you take your pulse to make it into your aerobic range. And you mustn't fall below twenty-two heartbeats at any time during a period of twenty minutes.

I rarely take my pulse because I know what my aerobic level feels like and I'm not trying to lose weight. If you don't want to get too scientific, just remember that aerobic exercise merely means working hard enough to mildly sweat and pant for twenty continuous minutes. Don't stop for more than a few seconds or your pulse rate will start to drop. Another approach to aerobics is to design your own "course," that is, you walk a convenient measured distance in a certain amount of time. My shortcut guide is to walk at what I know is a pace of three to four miles an hour.

However you organize it, to reach and maintain an aerobic level requires self-discipline and concentration on the goal. You can't just amble, stop to contemplate butterflies, window-shop, or visit with friends. But you don't have to drive yourself unmercifully, either. If you can't carry on a slightly breathless conversation with your companion, you're working too hard. But don't get *too* breezy. You *do* have to push yourself. Remember, speed and distance don't count. Only your age and your heart rate matter. As you grow fitter you'll have to go faster and farther to maintain an aerobic level. But that's a good sign. It *does* get easier and gradually becomes more fun.

When I'm in Hawaii, I often take an aerobic walk in the cool morning before breakfast and starting my day, unless I'm planning to hike, play tennis, or otherwise exercise later on. I look forward to my morning walk, whether alone or with my wife, for the good feelings it brings—mental, emotional, and physical. The mental and emotional benefits are the well-kept secrets of those who've become addicted to serious walking.

Mental & Emotional Benefits

Most of the runners and walkers I know agree that a large (but often unmentioned) component of the appeal of aerobic exercise comes from the mental escape, the opportunity for introspection and meditation, or the companionship afforded by dayhiking. My wife and I, while hiking, talk about things we otherwise never quite get around to, confident that we won't be interrupted by the telephone, children, chores, TV, or doorbell. We plan the future, we reveal our secret hopes, feelings, and aspirations. And we return refreshed, eager for the rest of the day.

When I'm alone, the comfortable rhythm of my feet on the trail seems to open my brain in a way that doesn't happen elsewhere. The privacy and freedom from interruption make for unhurried deliberation. Though I'm moving briskly, my thought processes are leisurely and deliberate. I can examine the subject at hand with maximum clarity in stress-free depth. Or enjoy the country as it passes me by. And I often return from my morning jaunt with a solution, plan, or insight that previously eluded me. I've come to rely on these solo morning walks for problem solving as well as aerobics.

Quickies

Remember, back in grade school, how you exploded from the classroom when the bell rang for recess, eager to run and jump and yell in the fresh air and sunshine for ten blessed minutes, gratefully releasing all the pent-up stress of hours of sitting and concentrating? That need, psychologists agree, isn't confined to kids, and it doesn't go away. It may get buried, but it actually grows stronger. The need for adults working indoors at desks all day long to relieve built-up stress is enormous.

People with accumulated job-related tensions will benefit just as much from a "recess" from labor as a classroom of kids after an hour of exams. That's why an increasing number of businesses, big and small, are giving their employees facilities for exercise on their lunch hours and coffee breaks.

Letting workers let off steam, they know, makes them more productive in the long run.

Time to Kill? Go for a Walk!

It's good business to build minigyms, rooftop running tracks, and par courses. Overall production is higher when workers have time off at crucial times throughout the pressure-cooker day.

Just because your company isn't yet in tune with the times doesn't mean you should spend your coffee breaks and lunch hours slouching in the lounge, drinking coffee, eating junk food, and breathing cigarette smoke. Instead, eat an orange and go for a brisk walk, even if you only have five minutes. You'll return refreshed and wide awake, with a clear head and a fresh spurt of energy. Take a break between tension-filled meetings or lengthy seminars at a conference to get outside, do some quick stretches, take a few deep breaths, then walk with all the gusto and vigor you can muster. It'll pick you up.

If your bus doesn't leave for five minutes, spend them walking instead of sitting. In fact, anytime you know you're stuck with a wait, instead of fretting indoors, go walking outside. If you "think walking" and look for opportunities, you'll be surprised at how many there are. Instead of sinking into an easy chair when you get home at night, flicking on the TV news, and reaching for a drink, put on a coat and go walking.

Wondering what's in the refrigerator to snack on? Escape temptation. Get yourself out the door and go walking. Whenever my attention begins to wander while I'm writing, I get away from the computer and step outside to draw a few deep breaths, stretch, and go for a walk. It never fails to refresh me, even if I only spend three minutes circling the house.

When I was a boy, cooped up perhaps on a rainy afternoon, my mother would insist that I "run around the block before dinner," no matter what the weather. It was sage advice. When I returned, I always felt more lively and alive, my spirits perceptibly lifted. A quick brisk walk is also an excellent way to overcome vague feelings of chilliness or depression. Whenever you've got a few spare minutes to kill, spontaneously go for a walk. Spur-of-the-moment and "opportunity" walks can lift you from the doldrums and turn your day around.

Half-Day Hikes

Sudden-inspiration walks can last an hour or more if you've got the time and need, but setting aside half a day for a hike usually involves some planning. Besides deciding where to go and making plans to reach your starting point, you'll probably make decisions about footwear, extra clothing, water, snack food, or a meal. To keep your hands free—a necessity, I believe—you'll probably need to carry a pack of some sort.

Seasoned hikers agree that it's generally a mistake to carry anything in your hands (except perhaps a walking stick) while hiking. By destroying your rhythm and drawing you off balance, the slightest burden in the hands will needlessly tire you out or produce an aching neck or back. It's almost as bad to carry anything in your pockets. Your trousers will bind on your legs or pull down uncomfortably on your waistband or belt.

If you carry a jackknife in your front pocket, you'll needlessly lift it every time you take a step. If it's hot, loaded pockets will make you hotter. I rarely carry anything more than a bandana in my pockets, and if it's hot I carry it in my pack. (The nylon running shorts and swimsuits I wear walking often have no pockets, removing the temptation to fill them.)

When you've got a whole half day free, you can get away for a *real* walk, maybe in the wilds if you're lucky. If you've got a half day, by planning ahead you can hopefully escape the worst vehicular traffic. If you can't get to the country, you can find a trail or fire road in a park or walkway. Take the trouble to find a safe, healthy place to walk. You deserve it. Give yourself a destination: a hilltop, restaurant, viewpoint, friend's house, ridge, or museum. Try if you can to return by a different route. Your outing will be more refreshing and enjoyable if you take a little trouble to plan it.

Beware Road Walking

Avoid walking along busy roads or highways. It's dangerous and unpleasant. Many an innocent walker has been hit. Road walking carries a high mortality rate! If you must walk along a busy roadway, remember the law says "keep to the left" in order to face oncoming traffic. At least you won't be hit from behind. Not only is there the menace of speeding cars and trucks on the shoulder of a road, the partly burned gasoline will poison your lungs. The noise and intimidation are almost as bad. A big part of the benefit of walking is the escape of indoor pollution, an escape into fresh country air. Roadside air is highly toxic.

I refuse to share my walking space with whizzing vehicles. Instead, I get in my car (or on my bicycle) and drive to a trailhead where I can set forth in peace, able to absorb the blessings of walking in the country. Generally speaking, the longer you've got to walk, the higher the quality of the environment you get to walk in. Planning ahead makes the difference.

All-Day Hikes

When I set aside the whole day for a hike, I make it an event, a special occasion, an adventure. I'm determined to go somewhere exciting. I'm either walking through beautiful country or going to a place I've never been. Maybe I'm taking an adventurous cross-country route to a known destination (like a lake, stream, or fishing hole). Often I'm climbing a peak, ridge, or hill.

There has to be something about the trip that's new and intriguing. If I can, I'll avoid retracing my steps. New country all the way keeps the trip interesting. I do my best to avoid the afternoon plod home on the same trail I hiked out that morning. Here are several useful strategies:

1. Plan to hike roughly in a circle. By starting at a trailhead from which several trails radiate, it is often possible to depart on one trail and return by another. Sometimes the adventure comes in figuring out a cross-country link between them.

2. Arrange for someone to meet you with a car at your destination at a prearranged time to pick you up and bring you home.

3. If you can't arrange a pickup, next best is to set up a shuttle in advance. You and your companions drive in two cars to the destination, leave one car behind and drive to the trailhead to set forth.

4. Divide the party in two, with half hiking the agreed-upon route in one direction, half going the other way. The two parties trade car keys when they meet on the trail halfway in between—perhaps for lunch. Friends of mine call it a "key hike."

Weekend Walks

An all-day hike may involve considerable travel to the trailhead if the goal is getting into wilderness. To get the most out of a weekend, your schedule might call for leaving right after work on Friday night, eating on the way, and staying in a motel, campground, or inn near the trailhead. When I was younger and employed full-time, that's how my friends and I began all of our summer weekend backpacks into the High Sierra. We didn't waste a minute. After sleeping at the trailhead Friday night, we would head up the trail after breakfast on Saturday morning—free until 8 A.M. on Monday.

My persistent memory of those weekends is delight at being back in the mountains. But I also remember the fatigue. We would drive from sea level to 8,000 or 9,000 feet. Hiking uphill from that elevation, carrying an unaccustomed pack, was hard work. By the time we reached our destination—say, eight miles away at over 10,000 feet—we were too tired and woozy from the altitude to do anything but make camp, cook dinner, and fall into the sack.

On Sunday, after breakfast and an hour of fishing, we took a last wistful look at the peaks, shouldered our packs, and headed back down the same old trail to the car for the long drive home. The trip goal (I see now) had been merely to spend a night in the high country, hoping for solitude. Often enough, we weren't even alone. There might be almost as many backpackers camped by our lake as there had been car campers at the trailhead!

Nowadays, my priorities are different so I set different goals. I want to enjoy as much country as possible in maximum comfort.

Instead of a weekend backpack, with all the meticulous planning, spartan meals, and heavy carrying, I'd plan to car camp in style and luxury both Friday and Saturday nights, taking dayhikes with only a light daypack on both Saturday and Sunday. I'd follow the wise admonition to "climb high, sleep low" to reduce the threat and stress of altitude sickness. Without a backpack to weigh me down, I'd be free to see almost twice as much country—and this time I'd be free to really enjoy it.

For instance, I could dayhike unencumbered to the same lake on Saturday and return that night to a gourmet dinner with beer or wine at the trailhead. But at the lake I'd have the energy to go fishing, do some off-trail exploring, maybe even climb a small peak. Or it might be possible to find my way cross-country over a ridge into the next watershed, allowing me to return to camp by another trail or down the ridge. And on Sunday I could wander in new country before returning to the car for the drive back home.

That's how dayhiking can be superior to backpacking—as long as the camping at the trailhead is pleasant and there are several appealing destinations to choose from. It's also a taste of my "extended" dayhikes, joyfully elaborated in chapter 7.

Combo Trips

The fifth type of dayhike is the combination trip—a dayhike/backpack hybrid—with the emphasis on dayhiking. Instead of just a weekend, let's suppose you have at least three days. Most backpackers would plan a three-day trip on which they'd hike as far as they could go, camp overnight, pack up and hike somewhere else and camp, pack up and come home, playing pack mule all three days.

Here's how I'd make that into a less arduous, more enjoyable three-day combination trip. First comes the planning. You need the right kind of country for this sort of trip, as you'll see. On day one I'd backpack into the wilderness by trail a short distance, leave the trail (if possible), and go a little way cross-country—as little as a quarter mile—to set up a base camp in wilder, less populated country. Then I'd spend the rest of the trip happily dayhiking, exploring, fishing, climbing peaks, or loafing, until I'd worn out the options or it was time to go home.

I'd try to minimize backpacking and trail travel and maximize unburdened dayhiking and cross-country exploring. I know this type of trip is not feasible everywhere. If you can't get off trail it doesn't work, unless the trailside camping is good and there are plenty of trails to explore. The combo trip is ideal for introducing newcomers to the wilds, for girlfriends and favorite uncles. It's even better for families with young children.

When she was eight years old, we took our daughter Angela on several backpacking trips whose distance from trailheads could be measured in yards, not miles. By scouting ahead of time and choosing the campsite carefully, we could provide her a full-fledged backpacking experience in a beautiful and solitary wilderness site.

She also experienced the achievement and satisfaction of carrying her pack all the way. After making her bed she had plenty of energy left for dayhiking nearby and feeding the squirrels before helping her mom cook dinner in the wilds. She still talks about those trips and asks for more exactly like them. They were easy on her parents, too, and fun, because the distances were short and the emphasis was on enjoying the wilds unburdened, not on carrying a pack for miles.

THREE

Secrets of Going Light

Meet Your Body's Needs . . . Clothing Design Theories . . . How Vapor Barriers Work . . . Free Vapor Barrier Socks . . . Chimney Venting . . . Fabrics & Coatings . . . Insulation Dynamics . . . Down, Pile, & Thinsulate . . . Foam Is King

It's important to be prepared, but carrying everything you could conceivably need takes all the fun out of dayhiking. The art of carefree walking is to take as little as possible, yet maintain minimum standards of comfort and safety. My rule of thumb is "skimp." After all, the dayhiker's play ethic is freedom: freedom to run or skip or climb boulders or trees. A big pack or heavy boots drag you down in more ways than one. The passport to dayhiking adventure is lightness—and that includes a light heart.

In Hawaii I often dayhike with a burly, sweating friend who carries twenty pounds to my three. And on extended dayhikes, I commonly see people laboring under bulging full-size backpacks. They lumber along, eyes on the trail, never stepping off it. They might as well be backpacking! To me, big loads destroy the lightness and freedom that are the special essence of dayhiking. So, I fight to stay light.

On quickie walks and aerobic workouts, I carry exactly nothing, and I wear the lightest clothing and shoes I can find. When you know you're going to sweat or at least warm up, deliberately underdress—providing you're going to keep moving. But don't risk becoming chilled. You have to make sensible preparations for any walk, even if it's only a glance out the window to check the weather.

For longer walks, anticipate possible weather changes and delays. Figure out the worst conditions you might encounter, plan for contingencies, then cut back as much as you dare. Don't rob your jaunts of fun and adventure by playing it too safe. Keep your dayhikes larks and you'll develop a lifelong addiction to walking.

Meet Your Body's Needs

The art of dayhiking is to carry everything you need for minimal comfort and safety and nothing more. You must avoid the risks of hypothermia, hunger, chilling, soaking, and so forth, without carrying an extra ounce. To scientifically figure out what you need in the way of clothing, it's first necessary to examine the body's needs, peculiarities, and sensitivities, then to look at the materials and strategies that govern the design of clothing to be considered. Only then can critical choices be wisely made.

The human body likes air temperature at the skin to stay about 75 degrees Fahrenheit, with a relative humidity of 70 to 95 percent. It tries to keep a quarter-inch cushion of moist, warm air against the skin. By maintaining this invisible layer of protection, the body can get by on as little as a pint of drinking water a day. Destroy this second skin with chilling temperatures or wind, and sweat glands will open wide, as the body goes all out to replace it by pumping heated water vapor through the skin.

But all parts of the body don't react alike. The body's thermostat is located in the chest. Inadvertently chill it and you trigger massive production of heat and moisture. When heat is being lost faster than it can be generated, the body constricts capillaries to cut down blood circulation. First to be cut off are the hands and feet, followed by the arms and legs.

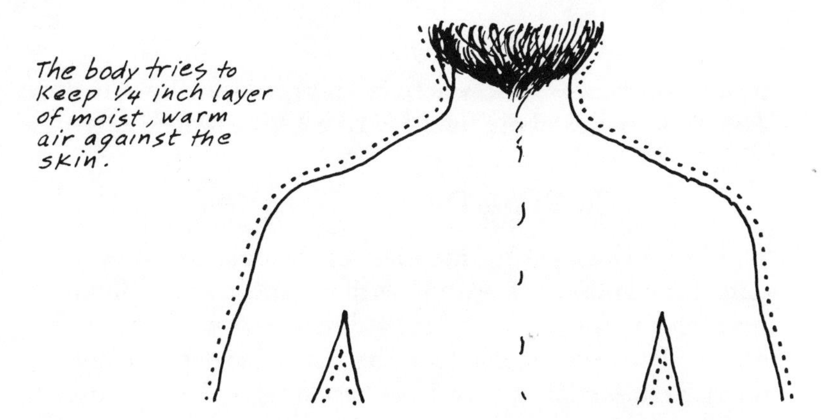

The head is the only part of the body in which the capillaries do not contract, because the body knows that survival depends on continued alert functioning of the brain. If the threat is great enough, the body can cut heat loss by a startling 74 percent.

There are two kinds of sweat, the one that gets you wet and "insensible perspiration," the one you don't feel because it evaporates as fast as it forms. But the resulting dehydration from both are signalled by sudden insatiable thirst.

The two greatest threats to body heat are wind (convection) and water (conduction) because each can swiftly wipe out the layer of moist warm air that protects the body. Water has twenty times the conductivity of dry air, and wet skin loses heat more than two hundred times as fast as dry skin. Combine evaporative and conductive heat, and it's clear why wind on wet skin is so devastating. It doesn't have to be cold for hypothermia to kill you. Most deaths occur at mild temperatures.

What does all this tell us about managing in the wilds? It warns us to cool off cautiously without triggering a deluge of perspiration, by ventilating our parkas at the waist, the cuffs, and the neck, not by baring our chests to cool wind. It tells us that putting on a warm hat is more important than putting on mittens when our hands get cold. It reminds us to cut down heat and moisture loss—as well as to drink liquids—when we get thirsty, because our bodies are signalling substantial dehydration.

When you want to cool off, use the power of evaporative and conductive cooling by soaking your hat or shirt in every stream you pass. Or, if water is short, repeatedly wet the back of the neck and the forehead like Arabs in the desert.

Clothing Design Theories

There are two opposing theories for keeping the body protected from cold while dealing with the inevitable buildup of unwanted body moisture: breathability and vapor barrier. I am proud to have introduced the vapor barrier concept to America's hikers in my two backpacking books. Because it works!

Breathability adherents fear condensation and claim that the body's water vapor—an unavoidable companion of body heat—will be satisfactorily dispersed away from the body without wetting clothing simply by putting porous fabrics close to the skin. It sounds okay, but in practice it doesn't work—not when it's really needed. When the body is working hard, it produces far too much water vapor for even the most porous of fabrics (like quarter-inch fishnet) to pass off.

That's why heavily advertised, expensive Gore-tex (and its imitators) doesn't work when you really need it. Gore-tex is merely a minimally porous coating glued to tightly woven fabric. Since it claims to be both wind- and waterproof, clearly the pores have to be microscopic. Obviously they can't pass a fraction of the water vapor generated by a hard-working body, especially if it's raining. But by far the worst aspect of the breathability theory is that it totally ignores the high convective and evaporative heat loss that inevitably accompanies water vapor production.

In stark contrast to the breathability myth, the vapor barrier concept takes clever advantage of the workings of the body. Instead of unrealistically trying to get rid of body moisture—while ignoring accompanying heat production—vapor barriers hold in body moisture and reduce its production, while effectively stopping heat loss. While breathability tries vainly to keep you dry at all costs, sacrificing heat along the way, vapor barriers keep your body warm and comfortably moist, while your clothes and insulation stay dry. They work with the body, not against it.

How Vapor Barriers Work

A vapor barrier is simply a sealed or waterproof fabric *worn close to the skin* to keep body moisture in instead of trying to drive it out. Don't groan at the thought of steamy streaming skin and soaked clothes. That needn't happen. Though sealed garments worn *away* from the body (like rain gear) can cause oceans of condensation, the same sealed fabrics worn *next* to the skin and kept warm produce a startlingly different effect.

This warm, sealed fabric protects and maintains the body's cherished moist layer of warm water vapor against the skin. Given the conditions it considers ideal, the body gratefully closes sweat glands and relaxes water vapor production—by up to 85 percent. The sealed fabric also protects the wearer's outer clothes from dampness, keeping them functionally effective and light. Best of all, the severe heat loss that always accompanies sweating is prevented, so the body's water needs and thirst are proportionally reduced. That means the dayhiker can carry less drinking water in arid regions.

Thanks to the retention of precious body heat, less heavy, bulky, expensive clothing is needed in cold weather, a boon to the dayhiker with only a small pack. The colder it is, the better vapor barriers work. And there's no drop in effectiveness when the weather turns wet and humid. Properly used vapor barriers will increase body warmth by a startling twenty degrees! I've bet my life on them many times in the last dozen years, and they've never let me down. To prove them to yourself, take the following simple test.

Free Vapor Barrier Socks

Take a large plastic bag from the produce section of your market and slip it over one bare foot when you get up some cold morning. Wrap the top around your ankle, put socks and shoes on both feet, and go walking in the snow. Not only will your plastic bagged foot stay dramatically warmer than the other one, but at the end of the day when you take off the bag, you'll find nothing but faint dampness—not the bagful

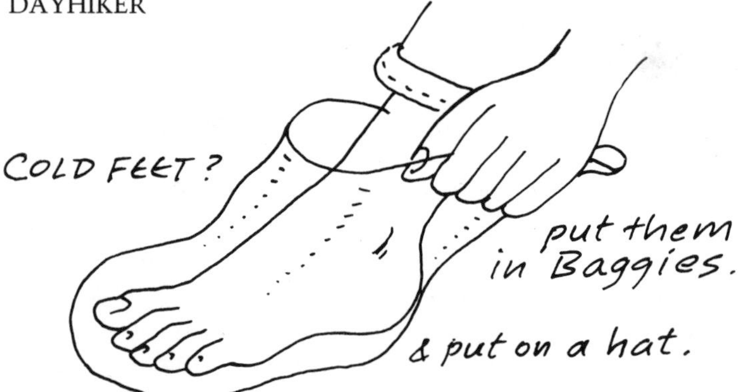

of sweat you expected. And the sock you wore over it will be totally clean and dry. Tomorrow you'll want bags for both feet!

What's the catch? The only drawbacks are: (1) Vapor barriers don't exactly feel like silk against the skin, but you can wear a thin layer beneath them. (2) They *can* produce heavy sweat if you—or the weather—get *too* warm. So save them for the cold. (3) And you have to stay active if you're going to depend on them entirely. If you plan to stand around in the snow for very long, you'll need some insulation over them.

For downhill and cross-country skiing and snowshoeing, I rely on vapor barriers exclusively, leaving home my parka, wearing nothing more than a light shirt and a wind shell over my vapor barrier (VB) shirt. And camping at the confluence of two glaciers at 15,000 feet in the Andes, I wore vapor barrier clothing from head to foot around the clock. I was always warmer than my mountaineering companions while wearing far less insulation.

Chimney Venting

When it's too warm for vapor barriers but not warm enough to go shirtless, I rely on "chimney" ventilation—not breathability—to get rid of unwanted body moisture and excess heat, especially if it's humid or raining. Instead of hoping that excess water vapor will somehow weave its way through successive layers of clothing, I take advantage of the fact that warm air rises. That's what makes smoke go up the chimney. If air is channeled to flow vertically upward through garments, cool air will enter at the bottom to replace the warm air escaping at the top.

To experience this effect and learn how to use it, put on a heavy shirt that fastens tightly at the throat and tuck it into your pants. Then exercise until you begin to sweat. Now untuck your shirttails and open your collar. You'll feel the warm air flow up around your throat as cool air is drawn up beneath your shirttails. In moments your torso will be cooler. More heat has escaped through chimney venting than could ever be dissipated through the most breathable of fabrics.

Chimney venting can be used to cool any kind of clothing. It works especially well with outer garments worn over a VB shirt, helping to prevent overheating. And it's effective under layers of ordinary clothing and beneath wind and rain gear. You can dial the degree of venting you want by controlling the various apertures in your clothing. Opening and closing cuffs and underarm zippers, if you have them, gives you added control. So does unbuttoning and rolling up sleeves.

The same principle, of course, works in reverse, if you want to get more warmth from your garments. If you're feeling chilly in a loose shirt, tuck it into your pants and/or button up the collar. When I choose my clothes, I consciously consider the venting options they offer, leaning toward garments with maximum range and flexibility. For instance, pullover shirts and shells make better vertical tunnels for venting purposes than do button-up or zip-up garments. Adjustable (snap, zipper, or button) cuffs at both wrists and ankles are more versatile than mere elastic.

Having taken a quick look at the body's needs and its strategies for self-protection, as well as at the principles

"Chimney venting" can prevent wetting from condensation.

designed to cope with them, it now ought to be easier to choose clothing that will serve you best on the trail. But first a word or two about the fabrics, coatings, and insulators from which they're constructed.

Fabrics & Coatings

Cotton is surely the most comfortable of fabrics against the skin—while it's dry. But when it's wet it's the worst. Furthermore, it's heavy and weak, tears easily, and rots and mildews when wet. Leave it home on serious walks, when your clothing and ounces count, and avoid it altogether in socks.

Wool is known for its springiness, resilience, and resistance to compaction, even when wet. It can absorb more water before saturation than most fabrics, but that isn't the virtue it's cracked up to be. Once wool is wet, it's heavy and hopelessly slow to dry—not to mention dangerously chilling to the body. As synthetics have improved, I find I use wool less and less.

Synthetics like nylon, Dacron, and Orlon have become the premier outdoor fabrics. They combine strength, toughness, comfort, light weight, elasticity, permeability, freedom from mildew, and other endearing qualities. There's now a supersynthetic for almost every purpose, putting natural fibers to shame. No longer do I sneer at synthetics. Some are as comfortable as cotton against the skin.

Fabric coatings are greatly misunderstood. Gore-tex, a film, not a fabric, simply cannot pass significant amounts of body-generated water vapor. Despite its well-advertised claims, it only passes about half as much moisture as the tightly woven nylon fabric to which it's glued. Therefore it isn't worth the substantial extra cost and weight.

Urethane coatings are commonly used for waterproofing garments. Two coats are better than one, but don't expect perfection even then. And remember that stitching will let moisture through any garment if not sealed. Honest urethane coatings make no claim of breathability. They are effective, inexpensive, and add little extra weight. I rely on them for my waterproof garments, counting on venting, not breathability, to disperse unwanted water vapor. Savvy clothing makers are increasingly designing garments for maximum venting options.

Insulation Dynamics

When comparing insulations, never forget that the body in motion makes any garment flex, creating a bellows effect. That means a certain amount of inadvertent venting of body heat can't help taking place, which greatly affects the theoretical capabilities of any insulator. So garment design is crucial—often more important than the type or thickness of the insulation used.

The purpose of any insulation is to hold in body heat. Though we talk of an insulator's comparative "warmth," it does not, of course, provide any heat. It only strives to conserve it, without getting soaked by perspiration—an almost impossible job. That's why vapor barriers are so important. They keep your insulation dry and effective and are far less vulnerable to ventilation losses.

Since motionless air effectively prevents conductive and convective heat movement, most lightweight insulation is built on the principle of small—quarter-inch or smaller—dead-air spaces. Exceptions to this rule are Thinsulate and closed-cell foams, which work on different principles. The effectiveness of other insulators are determined, at least theoretically, by their thermal conductivity, thickness, and uniformity.

Down, Pile, & Thinsulate

Down, once the standard of excellence in outdoor clothing insulation, has been losing its market in recent years to pile and other synthetics. Down is wonderfully light, but it's expensive, variable in quality, too easily compressed, useless when wet, and impossible to dry. Worst of all, however, is the near impossibility of building a uniformly insulated down garment. Its smooth bulk hides myriad gaps and cold spots. All these drawbacks, plus the demands of fashion, have turned outdoors people toward synthetics.

Pile has come a long way in a decade. Once a heavy, funny-looking fabric given to pilling, it's now the fashion leader in svelte, felt-like sweaters and coats in the latest decorator color schemes. Pile is still on the heavy side, but its ruggedness, insulating uniformity, relative inexpensiveness, resiliency when wet, and ease of drying have propelled it to the forefront among people who don't want to baby their insulation or worry about moisture. And it works well over vapor barriers.

Thinsulate claims nearly twice the insulating ability of conventional materials, thanks to a vacuum bottle effect that cuts conductivity. Heavily advertised like Gore-tex and fashionable like pile, its theoretical superiority as an insulator must be heavily discounted due to the bellows effect of unavoidable ventilation mentioned earlier. A vacuum bottle that leaks loses a lot of efficiency.

Foam Is King

Readers of my backpacking books may recall that I believe open-celled urethane foam is potentially the premier insulator for clothing, as well as sleeping bags, due to its combination of lightness, cheapness, uniformity, and resistance to wetting. But foam has never caught on, except among aficionados, due to its unavoidable bulk. Since most outdoor clothing is sold to fashion-conscious skiers and make-believe outdoorsmen, manufacturers design their slim and slinky products accordingly.

Closed-cell foams, which transmit no air and are therefore waterproof, have been more successful. Used in both wet suits (for diving and boating) and skiwear, these sleek, smooth foam garments—often one-piece suits—are stretchy, resilient, compression resistant, and tough. They offer vapor barrier warmth with built-in insulation, but they're heavy, hard to ventilate, and thus impractical for hiking in their present form. But that may change.

For greater details on body function, theories of design, fabrics, coatings, and insulators, see my backpacking books, *Pleasure Packing* and *The 2 Oz. Backpacker,* both from Ten Speed Press.

FOUR

What to Take

Featherweight Footwear . . . Socks . . . Shorts . . . Trousers . . . Vapor Barrier Protection . . . Shirts . . . Outer Gear . . . Headwear . . . Handgear . . . Survival Kit . . . Fannypacks & Daypacks . . . Beltpacks . . . Dayhiker's Checklist

Having investigated the body's quirks, demands, and reactions to the elements, as well as the materials and design strategies employed to combat them, let's now look at clothing, starting with the feet and working upward. Your aim is to get the most from what you carry and wear, to keep you safe and leave you free to enjoy an adventurous walk.

Featherweight Footwear

Since I don't like sweaty feet, sometimes I don't even wear shoes. For the past five years I've been taking lengthy walks in fairly rough country in winter Hawaii and the summer Sierra, wearing only sturdy Velcro-fastened sandals that lock down my heel. Originally made for river running by Teva and Sport, they are fast growing in popularity and are increasingly available wherever outdoor footgear is sold. I like them for beach walking or anywhere that I know my feet will get wet. And they're delightfully cool in the heat on easy terrain, where there's little or no menace from brush or falling rocks.

My first pair, which lasted nearly a thousand miles, had a toe thong, while my new ones have a strap that bridges my toes, permitting me to wear socks if it's cool. My Tevas provide admirable arch support, a quality rarely found in sandals of any kind. They weigh a hefty nineteen ounces and are built to last. But when I look down at the fresh puncture on my right foot, I'm reminded to warn you against wearing them off the pavement. Even on the best trails it's too easy to injure precious feet by accidentally kicking a sharp twig.

For dayhiking, some species of the myriad new featherweight walking, training, and running shoes are perfect. These sophisticated sneakers come in a plethora of subtle and cunning designs from a host of manufacturers. There are literally hundreds of models to choose from, so I won't make specific recommendations. But it's a pleasure to assure you that the masochism of heavy backpacking boots is now a thing of the past for carefree adventurous dayhikers.

All you have to do is show up at the shoe store—some of which now specialize in shoes for walkers—with the durable medium-weight socks you mean to wear on the trail and start trying on appropriate models. You don't have to understand all the sophisticated space-age materials employed. You don't even have to think about break-in or boot care. All you have to do is choose appropriate shoes that fit.

Appropriate doesn't mean old-fashioned sneakers. It means sturdy, substantial walking shoes with good arch support, padded lining, breathable uppers, and flexible, thick, cushiony soles. If they fit in the store they'll fit in the wilds. That's all there is to it. Well shod, you're ready to dayhike.

What a relief from the dark ages just a decade ago, when buying boots was a grim and nerve-wracking procedure. You bought expensive, heavy, leather iron maidens for torturing your feet. Some of those stiff, leaden foot-tormentors are still required for off-trail backpacking, but the emancipated dayhiker is forever free of those millstones around the feet.

The biggest advance over the all-leather boots I wore for decades is weight. Nothing's more important to a walker. Remember, one pound on the feet is equivalent to five—some say ten—on your back. Think what that means on a ten-mile walk. At two thousand steps per mile, it means you

lift five tons in ten miles for every pound of boot on your feet! Every pound you save conserves the energy needed to lift five tons.

So, don't saddle yourself with more shoe than you need. Low-cuts shouldn't weigh much more than a pound and a half a pair, and high-tops shouldn't go much more than two. Don't be influenced by the myth that high-tops are essential for ankle support. They're not. And don't even consider the lightest of leather boots unless you're sure the terrain demands it.

One of the great benefits of dayhiking is the freedom from stiff, heavy boots. On a given trail on a given day, a backpacker might need four-pound leather boots because of the weight he's carrying, while a dayhiker is appropriately shod to float up the trail in one-pound nylon running shoes. And don't rule out oxford-style walking shoes. They're making a comeback because shoemakers have learned how to build them out of weightless, high-tech synthetics that look heavy but aren't.

High-top basketball shoes and ordinary canvas tennis shoes are still better than boots or work shoes on the basis of weight. But decent arch support is vital in whatever shoes you choose. For snow, mud, or wet grass, I get added protection for my legs and socks by snapping on gaiters of waterproof nylon. But if it's really wet I don't expect my shoes or my feet to stay dry. Gaiters provide more protection to the sealed leather or rubber boots worn by backpackers and hunters.

Socks

Good socks, I believe, are as important as arch support. I choose mine carefully. A dayhiker depends on his feet to get him home, so he'd better take good care of them. Don't wear old, worn-out socks just because they're going to get dirty, and don't wear anything made of cotton. Once they get damp, cotton socks collapse, turning shapeless, soggy, and cold. You'll be sorry if you cheat on your feet by skimping on socks.

Socks serve four vital functions: they cushion the feet, absorb perspiration, reduce friction, and insulate against both heat and cold. For decades I chose wool socks for their springiness and warmth when damp. Now I rely largely on superior synthetics. Unlike wool, synthetics dry quickly and don't shrink when you throw them in the washing machine. The cushiony boot socks I buy nowadays are made of materials like Orlon acrylic, stretch nylon, Lycra spandex, and polypropylene.

Extra heavy socks can help make up for skimpy shoes, but remember, too thick a layer reduces shoe support and increases friction, while socks that are too thin can let you down in all four vital functions. With today's walking shoes or lightweight synthetic boots, a single medium-weight pair of genuine boot socks are all that most people need. But feet vary so widely, there's no single solution. People who don't walk much or have tender or irregular feet may find themselves susceptible to blisters or other foot problems when they start putting miles on their feet.

Of course, the problem may be the fit of their shoes. The solution may be to wear medium or heavy socks over an ultrathin pair of polypro inner socks. Some people lace their shoes tighter, or differently for each foot. The goal is reducing friction. Experiment a lot. Your feet are worth it.

When it's really cold, nothing will keep your feet warmer than vapor barrier socks, which are nothing more than plastic produce bags covering your feet beneath your regular socks. These bags will keep your feet toasty all day, while your socks stay dry and clean. Being warm, weightless, bulkless, and free, they're ideal for the dayhiker and perfect as backup or survival gear, too. I always keep a couple pair tucked in the far corners of my fannypack with the emergency wad of toilet paper.

You'll be surprised at how handy they are. If you don't have to wear them, you can use them for picking up trash on the homeward trail. When I'm relying on them for warmth, I take at least half a dozen because they're not very durable. When the first pair pops, I don't take them off. I just slip a second pair over the top, avoiding the shock of baring my warm, damp feet to the cold.

If you haven't yet accepted my challenge to test the dynamite power of vapor barriers by trying a bag on just one foot in the cold, here's a second chance. The colder it is, the more remarkable the result. They can save you from frostbite, but you don't have to go hiking in a blizzard to benefit. Elderly residents of a midwestern nursing home, who somehow got ahold of my backpacking books, wrote to tell me that plastic bags beneath their socks put a merciful end to their chronically cold feet. All winter long they faithfully wear them inside shoes and slippers. And at night they wear them under bed socks. Within a matter of days, everyone in the nursing home was wearing them.

The only drawback to VB socks is the slippery feel of plastic against your bare feet. But that can be avoided by sandwiching them between a light polypro inner sock and your regular outer sock. You may also have to lace your shoes a mite tighter to keep your slippery feet from moving in your shoes. Some people complain about the sudden chill that occurs when they finally take off the bags, but that only proves how well they were working. To avoid the shock, don't strip off your bags until you're ready to step into a hot shower.

Shorts

I do most of my dayhiking in shorts. I choose loose shorts or baggy swimsuits made of the lightest cottonlike synthetic material. I look for an elasticized or drawstring waist because I don't care for belts, and I demand front pockets, more to protect my hands than to carry unwanted cargo. I only wear a

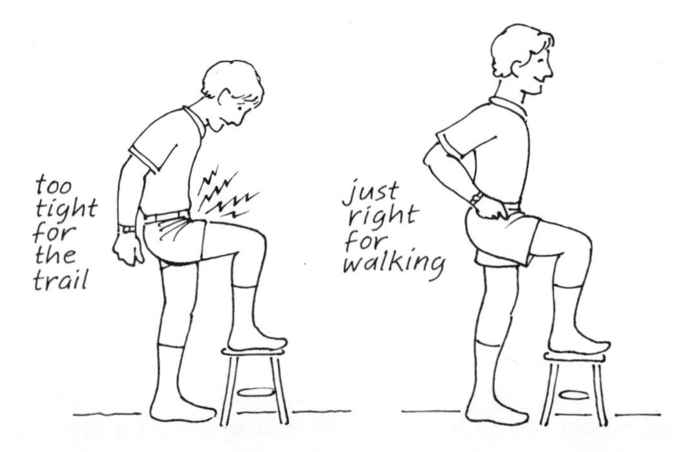

belt, usually of nylon webbing, when I need it to carry a beltpack, canteen, holstered knife, or the like.

The most important feature of any walking shorts is the freedom they provide to lift your knee to waist level without the shorts binding the top of your leg. That's my test. Fashionably slim or tight shorts are worthless for walking, no matter how sexy they look. The shorter the shorts, the easier it is to avoid binding. That's why I lean toward swimsuits with pockets. When shopping for shorts, be sure that the seams and hems in the crotch aren't bulky enough to cause chafing, or you'll be painfully raw after a couple of miles. By opening the outer seam on the hip, à la running shorts, you can stop the tendency to bind in your favorite shorts.

Pedal pushers, jams, and dress shorts are generally too tight for hiking. Flimsy nylon running shorts made of parachute cloth are unbeatable for comfort, especially in the heat, and I prefer them for aerobic walking despite the lack of pockets. Remember, if your shorts don't bag you're not in style for dayhiking.

Trousers

Unfortunately, the temperature and weather won't always permit shorts. If I feel I have to wear pants, I want the lightest pair I can get. My first choice is featherweight cotton pants with slash front pockets and an elasticized waistband. Mine weigh only nine ounces and roll up small enough to stuff in my fannypack when the weather warms enough, or I do. Under them I wear sheer nylon running shorts, so I'm ready to strip off my trousers in public and keep moving, rather than hunt for a private place to stop and change. With this combination, I'm also ready to go swimming.

If it's really cold and there's no possibility of switching to shorts, I may wear sweatpants over nylon boxer underpants. I want the loosest combination I can get, and sweats are dependably baggy—or should be. The next step up in warmth would be thicker pile pants. If still more warmth is needed, I wear eight-ounce polypro long johns or four-ounce tights underneath. Some people prefer panty hose or other stretch-elastic nylon garments. If it's bitterly cold, or windy or raining, I take along triple-threat vapor barrier rain pants.

No garment is more versatile than vapor barrier trousers, but you won't find them offered as such in the stores. Vapor barriers haven't caught on, because they aren't stylish and because breathability has been so heavily (and misleadingly) advertised, mostly by Gore-tex. But rain pants of various kinds make excellent VB trousers. Wear them against the skin, underneath any other trousers, and they'll keep your legs twenty degrees warmer. Two pairs, sandwiched between pile pants, will practically keep you warm standing still in a blizzard!

My favorite VB pants are urethane-coated nylon Sierra Designs rain pants that weigh eight ounces, with an elasticized waist and deep pockets. Drawstring elastic cuffs at the ankles permit good ventilation adjustment. I bought them for use as waterproof overpants for white-water rafting. They can double as rain gear if not needed for warmth, and they can triple as wind pants. So they're almost as valuable as a VB shirt for backup survival insurance for the dayhiker, because they're light, low on bulk, and inexpensive.

Vapor Barrier Protection

The cheapest and lightest VB pants are simply plastic trousers from a storm suit. They're low on fashion and durability, but they're terrific as multi-purpose emergency and survival gear. You may have to baby them to get them through a single trip, but they're an ideal first investment in VB clothing for walkers skeptical about the benefits I claim for vapor barriers. The top half will function as a VB shirt, and the suit will protect you from both wind and rain—if you're gentle with it.

Obviously, coated fabric garments are far more durable and dependable than mere plastic, and there are sealed fabric storm suits for as little as $25. Add plastic bags beneath your socks and disposable plastic gloves beneath your gloves or mittens and you've got the capability to warm yourself twenty degrees for pennies. And these low-bulk garments will fit in your daypack or fannypack, unlike the bulky insulated clothing you're used to lugging around.

The only drawback to VB pants is the lack of adequate ventilation in the crotch. To prevent heat buildup that could lead to condensation, a full zipper would be ideal. To experiment with ventilation I had a pair of ordinary cotton work pants altered by my friends at The North Face. The belt loops and button were replaced by a Velcro-adjustable elasticized waistband. And the fly zipper was replaced with a double-opening coil zipper that runs all the way from my front waistband, through my crotch, and up to my back waistband.

When I feel overheated, I simply open the zipper. Letting air flow through the crotch prevents, stops, and even dries up perspiration and condensation. And it's fun to watch the expressions on people's faces when they think they've found a man with his pants on backwards. They may look funny, but they work!

Shirts

Starting at skin level and working outward, there are two ways to layer garments over the torso. If it's really cold, I depend on a vapor barrier shirt. If it's too warm for VB but still somewhat cool, I'm obliged to use a series of ventilated conventional garments. Vapor barrier shirts and vests may be thought of as a substitute for a heavy, bulky jacket that wouldn't fit in a small daypack. Furthermore, these magic undershirts reduce the body's need for drinking water by reducing the level of perspiration, and they eliminate sweat in clothing.

Like VB pants, the VB shirt is a triple-threat garment, protecting against wind and rain as well as cold. But it's even more valuable than trousers because torso protection is essential to survival. All you have to do to get VB benefits is keep your VB shirt warm by wearing something over it. A light shirt will do. For wind and rain protection, you wear the shirt as an outer garment, employing ventilation to avoid condensation. Underarm zippers make the job easier.

If VB shirts have a drawback, aside from the feel of coated fabric on your skin, it's the need to develop acute awareness of your skin temperature when you wear them in less than freezing weather. If you overheat sufficiently before opening the shirt for ventilation, you'll find yourself suddenly wet with perspiration. Fortunately, it can't soak your nonabsorbent VB shirt or any of your protected outer garments, but it may force you to hastily undress for (usually) just a few seconds to allow your torso and shirt to dry. It takes practice to develop this needed awareness.

Happily, custom VB shirts can be purchased. You don't have to settle for rain gear. The shirts come with snap and Velcro front fastening, front zippers, or in a pullover style, with or without underarm zippers. Since they're worn against the skin, they need to fit snugly to avoid wrinkles and folds. But if they're too tight, you won't be able to wear them as outer gear over other garments for wind and rain protection. My collection of VB shirts weigh six to seven ounces each.

Any impermeable (i.e., sealed) upper-body garment can potentially be used as a vapor barrier shirt. Rain and storm suit tops don't work quite as well as the pants unless you buy them small. But you can make a dynamite VB vest that's almost weightless and practically free from a plastic garbage bag or clothes cleaners' bag. Turn it upside down and cut holes in each corner for your arms and one in the middle for your head. Now you're looking at a VB vest.

Put it on over your bare torso or over thin underwear and tuck it into your trousers. Then over it put on a relatively snug shirt or light sweater that will hold the vest close against your body. Your VB vest will keep you nearly as warm as a custom VB shirt or a parka. And it will wad up into a tiny, lightweight ball in the corner of your daypack to provide you insurance against a sudden change in the weather.

People who can't bear the feel of coated fabric or plastic on their skin can wear vapor barriers over a sheer open-weave undershirt. The ideal garment, far better than a T-shirt, is a thin, synthetic net undershirt with the largest possible holes to permit the free passage of moist air. Mine commonly weighs as little as two ounces, far less than the old Norwegian cotton fishnet or string shirts I used to wear. Wearing an undershirt cuts down the efficiency of a VB shirt slightly, making it more important to keep it warm and snug against the body.

When it's definitely too warm for a VB shirt (unless I want to risk oceans of perspiration), that sheer net shirt is the foundation layer of my layered clothing. It's probably the only garment that deserves the term "breathable." It isn't immune to condensation if conditions are bad enough, but it helps maintain the body's layer of moist warm air against the skin, while helping excess moisture escape. You avoid the wholesale evaporative chilling inflicted by a sweat-soaked cotton T-shirt.

Instead of open net, I sometimes opt for a long-sleeved polypro pullover. This nonabsorbent nylon undershirt covers my rear, and its turtleneck zips up beneath my ears. It weighs nine ounces and is comparatively bulky, but it's sure to keep me warm on those borderline days when it isn't quite cold enough to wear a VB shirt, because it conserves the heat in my arms and keeps me from losing it from my throat and neck.

On top of one of these two undershirts goes a conventional shirt, according to my needs. It may be a flimsy denim button-up work shirt worn primarily for mosquito or sunburn protection, but most often it's a Ben Davis long-sleeved ten-ounce Hickory pullover. Tucked in and zipped tight against my throat, it can provide considerable warmth. Because it's a pullover, it provides effective chimney venting when opened. If I'm not quite warm enough, I'll cover my overshirt with a seven-ounce wind shell, rather than put on a sweater or wool shirt.

If those three layers—closed tight to ventilation—aren't enough to keep me warm, I know it's cold enough to wear a VB shirt, which means I won't need a sweater or insulated jacket. I don't even own a big parka anymore. If I need insulation between my VB shirt and my outer parka shell, I wear a pile coat that covers my bottom. Beyond that, it's too cold for outdoor dayhiking—for me, at least.

Outer Gear

Over my outer shirt I usually wear an anorak: an unlined, pullover parka shell. Both my old uncoated wind shell and my new urethane-sealed Sierra Designs anorak weigh in at only seven ounces. I rarely go on a serious dayhike without one of these two garments in my pack. They're my outer line of defense and have saved me from hypothermia more than once. Since I rely on them for protection from the elements, their design is of utmost importance.

My old green wind shell of tightly woven nylon has a drawstring for tight closing at the hips, elastic cuffs, a roomy front hand-warmer patch pocket reached by two diagonal zippers, a zipper at the throat, and an integral hood with a drawstring that can close the opening for my face to the size of a baseball. I can wear it with the ventilation open for wind protection on a sunny day. Or I can close it down tight if a sudden change in weather brings dangerous wind and cold. It's the single most valuable garment I've ever owned—and I haven't seen its equal in the stores. It's my number one backup and survival protection while dayhiking.

My blue Sierra Designs urethane-sealed anorak is of almost identical design. It's longer, covering my bottom, and has an additional deep front pocket behind the hand-warmer pocket, reached by a horizontal midchest zipper. I carry it instead of the green wind shell if there's a real possibility of rain or snow—perhaps as a companion to its matching rain pants. It's as versatile in its way as my green wind shell since it can double for wind protection and triple as a vapor barrier shirt when worn against the skin under something else to keep it warm.

It's hard to find urethane-coated shells or parkas of efficient design because the public has been brainwashed to believe that rain gear must be breathable, since sealed garments just *have* to get you wet from condensation. Just the opposite is true. Gore-tex's breathability is so negligible under critical field conditions that it can't prevent condensation, while well-ventilated urethane-sealed rain gear beats condensation, especially when equipped with underarm zippers.

The public doesn't realize that *all* parkas perform well under ordinary conditions. There's no reason to pay staggering prices for heavy Gore-tex, when venting provides the answer to condensation. And it's foolish to carry big expensive bulky parkas—which defeat ventilation—when vapor barriers can cheaply provide the same warmth without bulk, eliminating sweat in clothing while reducing the body's water needs as well as the need for extra clothing.

Headwear

Veteran outdoor travelers know that hats are the first line of defense against killer hypothermia. When it's windy or cold, protecting the head from heat loss is vital because the capillaries in the head and neck don't shut down when chilled like those in the rest of the body. Heed the old adage, "If your feet get cold, put on a hat." The best all-around headwarmer is a bulky knit watch cap of wool or synthetic. It's the ear-covering fit, not the material, that makes it work. The same hat made of windproof foam would be far better—if there were one—because it would function as a vapor barrier.

Better than a watch cap if the chill factor is serious is a balaclava, which protects the neck and throat as well as the head, or a ski mask or cowl. Hats are also vital for protection against sunburn, glare, rain, wind, and bugs. The choices include shapeless felt hobo hats, which can be ventilated with a little judicious snipping, tennis hats, straw hats, caps, net-brimmed hats, and so forth—depending on their principal use and individual taste. Visors and net baseball caps with long bills are good for glare but don't protect the ears from sunburn. If wind is a threat, all but the snuggest headgear can be equipped with a pair of nylon ties that knot beneath the chin.

Handgear

If I carry a watch cap, I also take mittens. I prefer mine knitted from nonabsorbent acrylic rather than wettable wool. There are a great many weights and cuff lengths to choose from. It won't surprise you to read that any mittens or gloves get a tremendous—twenty-degree—boost in efficiency if you wear disposable polyethylene gloves underneath. You'll find them sold in drugstores, markets, and paint stores. Kept warm and loose, they function as efficient vapor barriers.

Usually one size fits all, but be sure they're not tight. It doesn't take much pressure on the skin to constrict blood flow in the hands, thus cutting down needed heat production. I take a pair or two whenever I carry gloves or mittens because they're weightless, low on bulk, cheap, and fragile, and they double the warmth in my hands.

Jack Stephenson, the father of vapor barriers and the man who introduced me to their warmth, used to spread the good word when downhill skiing by taking a pocketful of polyethylene gloves with him and handing them out to the people he met on the slopes and lifts. He gave them just one glove to try beneath whatever mittens or fancy ski gloves they were wearing, inviting them to see if they noticed a difference. Within an hour they were hunting for him to beg for a second glove.

Survival Kit

There are times—on serious trips in difficult country—when dayhikers should carry some kind of survival kit for emergencies or delays. Mine consists of an old tin cup, which can be used for cooking, a dozen kitchen matches, a yard and a half of toilet paper, fifty feet of nylon cord, two big plastic bags, a quarter of a rum-fudge bar broken into squares, three tea bags, three bouillon cubes, several sheets of folded paper, and a pencil stub. All this packets tightly into the cup, which is packed in one of the plastic bags. Total weight is only six ounces.

Other items that might be worth carrying are: waterproof matchbox, compass, whistle, flashlight, 56-by-84-inch Rescue Blanket, and tinder (in addition to the TP and notepaper). In mild weather, if adversity fails to strike, I can always use my kit to brew a cup of tea over a fire beneath a tree while I wait for an afternoon lightning storm to pass.

A bandana or handkerchief of some sort is so versatile that it's usually a mistake not to take one. It's the only item, except for lip salve, that I'll carry in my pocket. Bandanas are useful as washcloths, towels, neckerchiefs, headbands, head-

cloths, cool compresses, napkins, slings, even toilet paper and sanitary napkins in a pinch. They can easily be washed out after use and hung to dry from pocket or pack. A headband may be useful to keep sweat from your eyes.

There are a number of items that my wife and I share. More than one per party can be redundant. We usually have with us our homeopathic first aid kit (see chapter 8), a tiny large-bladed knife, lip salve, a yard of toilet paper, foot-sized plastic bags, twenty feet of high-test nylon twine, insect repellent, a garbage bag, lunch, and snack food. Sometimes we also have occasion to take a map, tiny compass, water bottles, a map magnifier (which doubles as a fire starter), dark glasses on retainer straps, a notebook and pencil, a tiny flashlight, camera, fishing equipment, dental floss, ID cards, paper money—most of the survival kit. It all depends on the trip. For instance, we always take a tiny flashlight on an after-dinner ramble in unfamiliar country.

Fannypacks & Daypacks

In warm weather I don't like straps over my shoulders or the sweat from a daypack in the middle of my often bare back, so I prefer a fannypack. Its center of gravity is closer to mine and it doesn't swing when I walk. The fannypack rests against my clothing and is well supported on my strong sacral promontory, so I hardly know it's there.

In winter or for a week's extended dayhikes, I'm happy with a teardrop daypack. Fannypacks are no fun when they're overloaded. My big Lowe fannypack has a capacity of almost two thousand cubic inches, enough for a spartan overnight trip, but if I'm carrying more than about five pounds, it bobs up and down with every step, destroying my rhythm.

A lightly loaded fannypack lets you float up the trail without a care. I can even run if I want to. If I need to get into it, I just lift it to loosen the belt up and swivel it around to the front. Some people wear two of them, one in front and one in back, rather than submit to shoulder straps. But if I have to carry more than a single quart of water, I either hang it on my belt or reluctantly decide to take a daypack instead. Two quarts is just too heavy for any fannypack.

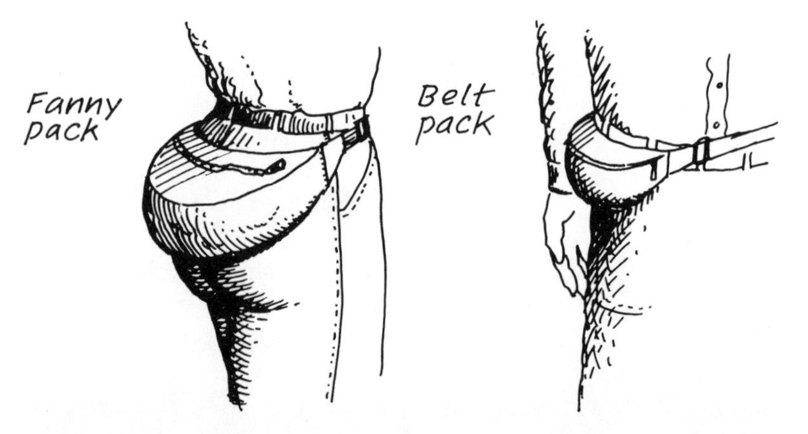

Like most other packs, fannypacks should be loaded with the heavy, dense items up against the body to keep them from bobbing. The best designs hug the back and are vertically, not horizontally, deep. Teardrop daypacks should also be flat against the back and no more than three or four inches thick, to keep the weight close to the body. Beware large back pockets which, when loaded, pull you backward. If you need more capacity, consider a rectangular knapsack design or climber's pack that covers more of your back rather than a thicker daypack. A flat profile is always best.

Daypacks can be extremely uncomfortable if they hit you wrong. Try them loaded in the store before you buy, checking sway, swing, pullback, buckle release, and strap adjustability. I want well-padded shoulder straps, a carrying loop, top outside attachment points, reinforced stress points, quality sewing, and well-covered zippers—features you may not find in supermarket offerings. A rugged daypack of coated nylon should last you half a lifetime, so be sure it suits your purpose and fits. My North Face teardrop pack is still going strong after more than twenty years.

Beltpacks

Beltpacks come in two distinct styles. There are coated nylon pouches that you thread onto a conventional belt and miniature fannypacks with integral belts that you wear in front or back. The largest of the latter are built for camera gear and may also be supported by a strap around the neck.

But they're good for a lot more than cameras. The smallest are oversize money-belts, or wallets or purses with built-in belts. All can be extremely useful to the dayhiker. Aerobic walkers and runners like the small ones because, when lightly loaded, they move with the body instead of bouncing, even when you're running.

I have a pair of individual pouches that I can mount on the front of any fannypack, giving it added capacity for small items, placing them where I can reach them in a hurry, and freeing up the fannypack for bulkier, heavier items such as water, lunch, and clothing. I use the pouches for carrying a small camera when I'm looking for pictures; lip ice; a map and compass if I'm groping my way through new country; a notebook and pencil if I've things to write down; a knife to sharpen the pencil; snack food if I'm hungry; a handkerchief if the cold is making my eyes water; money; ID—anything small that I may need fairly often but don't want to put in my trouser pockets.

Dayhiker's Checklist

In order to get ready to go walking in a hurry, without forgetting anything vital, I find it helpful to have a checklist of likely gear and clothing handy, so I'm ready for a spontaneous trip. It's also useful when pondering what to take on a serious exploration, summit attempt, or extended dayhike. Here's my checklist, with the weight of each item in ounces, to get you started making your own.

CHECKLIST

Packs	Weight in Ounces
Lowe Walkabout	40
Lowe (expedition) fannypack	20
Standard fannypack	5
North Face teardrop daypack	8
Beltpacks	2 (each)

Clothing	Weight in Ounces
Hi-Tec Sierra Lite boots	30
Teva sandals	19
zoris (as camp boots)	5–7
Nike running shoes	19
synthetic boot socks (pair)	3
VB socks (plastic bags)	–
VB gloves (thin plastic gloves)	–
lightweight liner socks	1
lightweight trousers	9
walking shorts	4
nylon running shorts	2
net T-shirt or net polo shirt	3
Hickory shirt	10
denim long-sleeved shirt	8
VB shirt	6
REI VB pants	8
polypro long-sleeved shirt	9
zipper-arm pile sweater	21
long johns	8
red tights	4
green wind shell	7
Sierra Design coated anorak	8
Sierra Design rain pants or VB pants	6
VB storm suit	21

CHECKLIST

Clothing, *continued*	Weight in Ounces
watch caps	2–3
balaclava	3
guide's or straw cap or tennis hat	2
big bandana	1
light mittens	1
heavy mittens	2
gaiters	4

Miscellaneous Gear	Weight in Ounces
homeopathic first aid kits	2–3
first aid supplies	3–4
notebook and pencil	1
nylon line	1
Swiss army knife	1
lip salve	1
insect repellent	2–3
#15 sunburn cream	1–2
fly rod, reel, line	8
Olympus 35mm camera	12
dark glasses on keeper	2
Mallory flashlight	3
lithium flashlight	1
plastic bags, garbage bag	1
toilet paper and wooden matches	1
compass and map	2–3
magnifier	–
canteen	5
survival kit	6
water bottle	4

FIVE
Where to Go

Discovering the Great Basin . . . Walking Wild Hawaii . . . California's Low Sierra . . . Finding Your Own Dayhikes

Finding great dayhikes is a lot like finding gold and often just as rewarding. Like the search for gold, it's an art as well as a science. Good places to dayhike are where you find them. There's no such thing as a dayhiking directory. Instead of trying to make one, I'm going to hope that personal examples will be more helpful.

In the past six years, I've moved my home to three distinctly different regions, each one totally new to me. In each new locale I was determined to sleuth out the best dayhiking to be found, because that's a vital part of my life. I plan to show in some detail the stratagems I employed to find a wealth of satisfying places in which to indulge my passion for daily walks. From these accounts you will hopefully be able to concoct your own battleplan for finding the very best walking in your area.

First, I moved from the city to the wild high deserts and ranges of the Great Basin, principally in Nevada. Next came the big island of Hawaii with its volcanos, jungles, and beaches. Finally, my bride and I moved to the high westside foothills of the California Sierra Nevada. All the while, I maintained a cabin above Lake Tahoe on the edge of Desolation Wilderness.

Discovering the Great Basin

I've always been attracted to the dry high desert just east of the steep escarpment of the Sierra Nevada. So finally, I bought land in a little mile-high valley on the wild West Walker River and built a retirement home near the village of Walker. When my wife and I unexpectedly divorced, suddenly I found myself living there alone, a stranger from the city, wondering where to go walking.

Usually the best place to start is with information officers at state, federal, and local parks and forests. They get paid to supply maps and point out trails to the public, but I was far from any parks or rangers. I was in nearly virgin country, so I had to rely on what the locals would tell me, plus whatever books and maps I could find.

Pestering my immediate neighbors for advice, I found out about a series of beaver dams up Deep Creek. When I went there I found the beaver were gone but brook trout remained in the ponds. On the way I found a road—actually a stock driveway—that led into the Sweetwater Mountains to the east. I followed the road to its end at nearly 10,000 feet, at the base of intriguing peaks. To find out what lay beyond, I ordered all the topo maps of the region from the U.S. Geologic Survey. Studying them I discovered ghost towns and pack trails leading to abandoned mines.

Armed with my topos, I returned to the road end and dayhiked up washed-out mine roads on the way to the summit of Mt. Patterson, highest point in the range at 11,600 feet. From the top, I could look across my valley at the glaciers of the High Sierra. Before coming down, I studied the terrain and, with the aid of my maps, planned a series of likely trips. The next time I was in a bookstore, I found a booklet on the region's ghost towns and old mines. I also found *Hiking the Great Basin,* a Sierra Club totebook by John Hart. It's a treasure. Studying the two books, I compiled a list of intriguing destinations.

Before long, I knew the western side of the Sweetwaters fairly well. And I had them to myself, except in deer season. I wanted to explore the range's steep east side, but the locked gates of a sprawling cattle ranch blocked access at low elevations. Before I could ask the ranch for permission to enter, I met an old-time well driller named Ed. (Whenever I meet a longtime local in good country I always ask a lot of questions.) With Ed I struck pay dirt. A mine operator had hired him to sink test holes high on the east side in search of a lost vein. When he saw me salivate, he invited me to go with him on his next trip.

The following week at dawn we ascended the mountain in his pickup truck through a series of locked gates to the diggings. There he let me out, offered me his pistol for protection, and warned me to be back by 5:00 P.M. sharp, when he'd be heading down the mountain—with me or without. With map and compass I set off to visit mines and climb Wheeler Peak, second highest in the range.

Coming down from the summit, I visited an abandoned mine. The miners seemed to have left in a hurry half a century before, perhaps when a gullywasher tore out the road. Jeans and jackets still hung on the wall. The kitchen table was set for dinner, and there was food in jars on the shelves. The toolshed was fully stocked, and fresh water stood in an artesian well just inches below the floor. I barely made it back to Ed's mine shaft by quitting time. I wonder how long he would have waited!

When I first moved from the city, I had transferred to the Great Basin chapter of the Sierra Club. When I received the first issue of the chapter's newsletter, *Toiyabe Trails,* I devoured its listings of local hikes. The Great Basin is so vast that some of these outings were four or five hundred miles away. Though I didn't go on many organized trips, I learned about a lot of nifty places by calling the leaders and asking a lot of questions. Some of these places I later visited on my own or with my new wife. (By now I had met and married a desert girl named Deanne, who liked to hike as much as I did.)

For instance, Arc Dome, the highest peak in the Toiyabe Range at 11,800 feet, had always drawn me. I'd read about it in both Hart's guide and *Toiyabe Trails.* One weekend, the Great Basin chapter scheduled a climb there, approaching from a low eastside trailhead. Their route meant a climb of nearly 5,000 feet, so I didn't see how they'd make the summit. Deanne and I decided to approach Arc Dome from the west the same weekend, saving a hundred miles of driving and starting 2,000 feet higher. We met their strongest hikers on the 11,000-foot summit plateau. We were on our way down, and they were exhausted and about to turn back.

I have also been on Sierra Club chapter trips to enchanting hidden places I otherwise would never have known. On one memorable three-day weekend in May, for instance, a group of us launched inflatable kayaks in a tiny seasonal river named the Quinn in northwestern Nevada, to paddle more than fifty miles downstream into the inaccessible heart of the forbidding Black Rock Desert, a last refuge of desert Indians. When the river sank into the sand, we let the air out of our boats, lashed them to our backpacks, and hiked out to the cars we had left on the edge of the desert. My only cost on this trip was a few dollars for food and my share of the gas.

I then found another guidebook, *The Mammoth Lakes Sierra* by Genny Schumacher (now Smith). It turned my attention from the empty desert ranges to the comparatively lush east side Sierra, which rose directly west of our river. The old-timer who built my home on the river had lived in the valley all his life. As a young man, Bill had been a ranger for the Forest Service and had spent a lot of time in the Sierra high country.

On my topo map he proudly showed me a lake that had been named for his wife, Emma. He said it was full of brook trout and he told me how to get there. It was an easy hour's dayhike after a forty-minute drive, so I went there often—whenever I grew weary of desert sand and wanted to see something green. And I never failed to catch hungry brook trout, though once in October I had to cast my fly through holes in the ice.

Exploring a dirt road up a Sierra canyon one day, I met a cowboy and asked him where it led. He told me about the remnants of a ghost town and a nearly intact stamp mill used for crushing high-grade gold ore—one of the best examples I've ever seen. On a lonely road, far back in the desert hills, I was stopped by an eccentric but entertaining hermit with a rifle and persuaded to browse his permanent yard sale. He told me fantastic tales about wolverines, claim jumpers, and desperados, and about the location of an old mine road that led to the base of peaks I'd always wanted to climb but didn't know how to reach.

The Nevada desert is studded with little-known, well-watered mountain ranges inhabited by deer and trout and surrounded by sand. These Shangri-las offer some of the best exploring in America. From my maps and books and Sierra Club periodicals I discovered and visited a number of them: Ruby Mountains, Shell Creek Range, Toquimas, Monitors, Duckwater. Many have since become official wilderness areas, and one is now a national park, so they are easier now to find but still mostly wild.

The rancher to the east of me turned out to be the owner of Mammoth Lakes Pack Station, so it wasn't long before he had me on a pack trip over Duck Pass into the Golden Trout country of the John Muir Trail. The only catch was that I had to ride a horse or walk in horse manure. But once we pitched camp and the horses were tied up, I was free to explore the crags and tarns on foot, a dayhiker once again. According to my topo maps, there was a string of little valleys with intriguing names—Indian, Fish, and Piñon—just east of where I lived, but the maps showed neither roads nor trails. How to get there? I took the only road leading in that direction, coming at dusk to a house in the mouth of a narrow canyon. The owner took me up the canyon and showed me an ancient trail that he told me would lead me to Indian Valley, where the local Paiutes still occasionally went to harvest piñon nuts in the fall.

Several days later I found my way into the valley and began a most rewarding exploration, the climax of which was a dawn-to-dark traverse of all three valleys, returning via a shuttle. In the course of my rambles I discovered the spring the Indians had apparently relied on, several of their encampments, obsidian arrowheads, tepee rings, and a mortar and pestle used to crush the piñon nuts into a highly nutritious mash. I went back recently, and it's all still there.

Before I left the region, I learned of a number of sparkling hot springs from the Indians with whom I played basketball. And the teachers with whom my wife and I played volleyball at the high school told us about their favorite trails and picnic spots. The Sierra Club's national magazine was responsible for our taking a six-day raft trip through the beautiful desert canyon of the Owyhee River. And a book called *Driving Back Roads* led us to still more wild and lovely places.

I only lived year-round on the river for three years, but in that time I found ghost towns, hidden springs, Indian camps, stamp mills, beaver dams, eagles, wildcats, trout streams, lost mines, and ancient bristlecone pines. By the time I moved to the other side of the mountains, I knew the country well enough to be regarded as an authority on dayhiking in the region because I had taken the trouble to look for good places to walk. You can do just as well in the country where you live or where you want to walk.

Walking Wild Hawaii

Five years ago my wife and I bought a modest bungalow in a village on the dry west shore of the Big Island of Hawaii and started spending our winters there. Deanne taught grammar school, I wrote, and the rest of the time we explored the strange (to us) terrain on dayhikes. As we had in the desert, we started from scratch. Hawaii Volcanoes National Park, with its continuously erupting volcano Kilauea, was the logical place to begin. Since the park was a three-hour drive across the island, we booked lodgings in rustic Volcano House, perched on the crater's rim, a short walk from park headquarters.

On our arrival we picked up maps showing all the park's trails, and that night we watched the hotel's spectacular movie of eruptions past and present. The next morning we paid a visit to the park's visitor information center, asked a few questions about conditions on the trails: likely temperatures, shade, drinking water, underbrush, and the grades to be encountered. Then we loaded fannypacks with necessities and set forth into the steaming caldera in shorts. That began our dayhiking in Hawaii. We're still exploring remote corners of the park.

Deanne discovered the Kona Hiking Club in a brief announcement in the newspaper. It invited the public to go walking on a roadless stretch of coast thirty miles to the south of our bungalow. She called for directions, and early in the morning we met a group of about thirty island residents at the end of a steep dirt road by the sea. As we walked through the jungle we made a point of introducing ourselves to various friendly looking people and asking them to tell us about their favorite island walks.

Over lunch, on the beach that was the hike's destination, we got acquainted with the club's leaders, Hank and Claire Swann, and learned about their forthcoming outings. That was the start of an enduring friendship. We still go walking with the Swanns when they visit our favorite jaunts.

We found the local (Moku Loa) Sierra Club chapter across the island in Hilo, and by phone we arranged to be put on the mailing list for its quarterly list of outings. As soon as it arrived we started calling trip leaders about the upcoming hikes we found appealing. Over the years we've climbed mountains, examined ancient temples, visited waterfalls, gone botanizing and birding, explored abandoned villages, discovered hidden ponds, walked miles of deserted coast, and probed deep jungles with the Sierra Club and the Kona Hiking Club, sometimes with local geologists, ornithologists, volcanologists, and botanists.

But we soon learned that much of Hawaii's loveliest wilds are privately or municipally owned and often officially closed to hiking. An acquaintance named Vic offered to take me hiking along a series of canals that fed his mountain town's reservoirs. For liability reasons, the water company refused to grant entry permits to hiking clubs, but they rarely hassled individual walkers.

On several of these walks Vic brought along a friend, Brian Nelson, who owned Waipio Valley Shuttle, a tour company that takes passengers in Land Rovers into the exotic jungle valley beyond the roads. Brian not only took us on the shuttle tour, he obtained keys to private gates beyond the valley and took us hiking back into the wild lands that lay between his valley and the watershed lands we had visited. In this way, via dayhikes, we gradually became acquainted with much of the wild country in the Kohala Mountains.

Vic had formed a unit of the ski patrol on Mauna Kea, the 13,796-foot dormant volcano that astronomers regard as the best viewing site in the world. In his four-wheel-drive truck he took me to the top. Then we hiked to the caves at over 12,000 feet, where the ancient Hawaiians came to quarry the hard rock for their axe heads and other tools. Blanks and broken axe heads still litter the area. We finished our breathless hike in the thin air at a high alpine lake, then bounced down the mountain to spend the last of the afternoon on the beach. Vic also took me sea kayaking until Deanne and I bought kayaks of our own.

The prettiest place Vic ever took us was Luahiniwai, the King's Bath, an idyllic pool of crystal-clear water surrounded by graceful palms, just behind an ocean beach on a roadless stretch of coast. We approached it on the King's Trail, an ancient groove through the razor-sharp lava that is studded with gleaming white stepping stones of water-smoothed coral. Luahiniwai quickly became our favorite destination, but unfortunately it is now privately owned.

A helicopter pilot who flew tours over erupting Kilauea told us how to safely get near the flowing lava. On the way, we stopped at the Queen's Bath, another lovely pool, but accessible to everyone by a short path from the road. We swam in its jungle-shaded waters just once. A week later, advancing molten lava from Kilauea buried it, burning the surrounding jungle and leaving no trace of the pool. The Queen's Bath, like the King's, is just a memory for us now.

When a favorite hiking partner from California arrived for a November visit, we decided to climb the island's other big mountain, 13,333-foot Mauna Loa, as a dayhike. Starting at sea level after breakfast, we drove to the end of the road at the observatory at 11,000 feet and started hiking—in shorts. It was cold but clear, and we gasped for breath in the thin air. In early afternoon we were suddenly enveloped by thick cold fog. We reached 13,000 feet before the cold and the lateness of the day forced us to turn back. We had started too late and weren't dressed warmly enough to make the summit and return.

To complete my tour of the island's four mountain ranges, Brian Nelson and I made a dayhike ascent of 8,275-foot Hualalai with the Sierra Club, passing through a cloud bank and emerging into the brilliant sunlight above. Our only companions above the endless sea of white were the island summits of Mauna Loa, Mauna Kea, and Haleakala on Maui.

By the time Deanne and I bought a Wilderness Press guidebook to hiking in the islands, we had already hiked all the trails on our island. But the trips it described on other islands intrigued us. At times, across the channel and above the mists, we could see the 10,000-foot summit of Haleakala Volcano. Finally we had to go there and explore its desert crater. And eventually we walked a lot of the trails on Kauai, Molokai, and Lanai as well. By ferreting out information from a variety of sources, we have found dayhiking trails to many of the islands' lovely wild places.

California's Low Sierra

When Deanne and I left the Great Basin, looking for a slightly less remote place to live, good dayhiking potential was a high priority. So was proximity to our mountain summer cabin on Echo Lake, accessible solely by boat or trail, at the edge of Desolation Wilderness. We found what we wanted in the forest near the mountain village of Pollock Pines. The old house needed work, but its two acres of big trees shaded a tiny trout stream. The place seemed remote but was only three minutes from the freeway and a supermarket on a road that dead-ended at the American River, half a mile below a small park and lake.

We switched our Sierra Club local affiliation to the Mother Lode chapter, while retaining our subscriptions to the Hawaii and Great Basin newsletters. Before long we received not only our new chapter paper, *Bonanza,* but also the newsletter of the Maidu subchapter in Pollock Pines. They offered listings of three or four juicy hikes each week, rated according to distance, elevation gain, and experience needed. In addition, there was free instruction in skiing, kayaking, canoeing, and mountain climbing. Chapter activities included dinners, slide shows, dances, movies, conservation projects, and work or cleanup trips to help repair and maintain the wilderness. Calls to the leaders of intriguing trips soon produced a growing list of appealing destinations.

With our canoe on top of the car, we drove to the river to explore Brush Creek and Slab Creek reservoirs, and dayhike up the feeder creeks.

We also started walking the forest lanes we could reach from our front door. Our neighbor, Ray, turned out to be a marvelous resource. A local history buff and walker, he directed us to faint roads and logging rails, old mills and springs. He even took us walking on trails he had built himself to complete a network of forest paths near our homes. And he showed us maps and books to document his tales. Before long we had developed so many routes, all starting at our door, that we could take a different walk every day of the week without repeating ourselves.

For instance, in less than ten minutes we can climb into a wild watershed with the rumbling cataract of the American River far below us, the snowy Sierra crest on the skyline, and unbroken forest in between—without a sign of the work of man. The path along the ridge to the east Ray calls "the Dogwood Trail." On cool days we follow the shadeless pipeline to where it plunges into penstocks diving back to the river. On warm days we walk the shaded trail around the lake or go up to the village for lunch. If it's hot, we follow the path beside the canal dug by the Chinese a century ago in the deep shade of silent forest.

To augment these delights, we paid calls on local Forest Service and chamber of commerce information offices for maps and information, coming away with printed matter on trail systems and networks we hadn't heard of. And at the library we found both books and magazines that yielded information on local walkways. The magazine rack at our market offered a wide selection of mostly national publications, like *Outside* and *Backpacker,* which often describe California Sierra trails of unusual interest.

The dam tender at the reservoir told us of a trail that climbs from the dam to our daughter's elementary school. We showed up at school and walked her home on a series of paths that beat the school bus home. After that she got the neighbor girls to walk with her. The school adopted another trail that had been trashed by loggers and helped restore it.

Deanne and I keep in shape walking every morning in our woods, but when we want a workout or crave long vistas, we head for timberline at the nearby Sierra crest. A forty-five-minute drive brings us to the trailheads at Wrights Lake on the west side of Desolation Wilderness, to the open expanses of Carson Pass, or to our Echo Lake cabin and its eastside trailheads. Having written the original guidebook to Desolation Wilderness, I know its trails well—almost too well. So when we yearn for the stimulation of something new, we fall back on the dozens of hikes offered every month in our local Sierra Club newsletters.

We also keep an eye on *Toiyabe Trails* for irresistible journeys to favorite haunts or places we've always wanted to see but didn't know exactly how to reach. We show up for chapter hikes whenever we can. For instance, recently we noted a listing for a Great Basin climb to one of our favorite summits, Echo Peak. So we climbed it from the opposite side, met the group for lunch on top, and showed them a new route down. Another day, instead of falling in line for a walk we knew well in Desolation Wilderness, we got a later start and caught up with the group at the destination, Smith Lake, in time to get acquainted over lunch. Then we climbed to a high pass for the view. Below us a coyote traversed the granite slabs, while an eagle circled above.

On the descent we again caught up with the group to visit in the shade by a spring. The trip provided us all the benefits of a group walk with none of the restrictions. We made new friends, avoided following the leader to a place we knew well, were able to hike at our own brisk pace, and had time left over to visit on the way home at the cabin of friends.

Several weeks later, however, we gratefully followed the leader's directions and chatted with friends we had made on the Smith Lake jaunt as we climbed 10,000-foot Red Lake Peak on a chapter trip into country that was new to us near Carson Pass. We ate lunch on the summit, savoring the panoramic view and the drama of a passing rainstorm. To parlay our foothold into future satisfying dayhikes, we returned on another day to scout the Carson Pass region by car, locating trailheads, walking around lakes, and sampling trails, so we'd be able to knowledgeably judge the merits of future Sierra

Club trips in the area. In this manner, we stretch our boundaries and add new country to our turf.

Finding Your Own Dayhikes

Hopefully, by now you will have noted the applicability of some of our strategies to finding satisfying dayhikes in your area. Don't despair because you live in the city. Cities provide more parks and walkways, and the walking is far better organized. Instead of begging gruff old-timers for directions, you can find your walks by telephone and obtain printed listings in the mail. You've also got much better bookstores, magazine stands, and libraries to browse than we do out here in the country, not to mention computer networks.

In the cities there are many more hiking clubs to choose from, even college courses to help you. But you still need to ask absolutely everybody who might conceivably help you for advice, favorite paths, and suggestions. And I urge you to join the Sierra Club in your area. Membership dues are $33 a year and include both national and chapter publications, which list hundreds of outings annually. It's an unbeatable way to get acquainted with new country and like-minded people, and chapter trips are free.

Or you can start your own hiking group. All it takes is a few friends who want to take a walk together on a semiregular basis. Scheduling a walk makes it happen—by overcoming procrastination and preventing postponement. Good intentions often aren't quite enough to overcome the inertia of busy lives. We sometimes go on Sierra Club outings simply because they're scheduled. Scheduling gets them on our calendars. We do what our calendars tell us! If you're at all like us, schedule dayhike, and you'll go.

In between walks, seek outdoor people, read about hiking, join outdoor groups, scout likely terrain, get your gear ready, savor your pictures, and dream over your maps. With enthusiasm, determination, and a venturesome attitude, you can't help but find the gold of richly rewarding dayhikes in the territory you call home and the country you visit on business and vacation. For ways to discover the best in extended trips—here and in foreign lands—see chapter 7.

SIX

Food for New Fitness

Big Breakfasts Are Bad . . . I Come On Board . . .
Experimenting with Fruit . . . How Our Bodies Work . . .
Eating & Exercising Strategies . . . The Rest of the Program . . .
Hard-to-Digest Foods . . . Essential Liquids . . . A Menu for Dayhiking . . .
Smart Trip Planning . . . Packing the Right Lunch . . . Treating Suspect Drinking Water . . . After Fruit Comes Lunch . . . Testing the Lunch Menu . . . "All Fruit" Days . . . Conduct Your Own Test . . . Evaluating Traditional Trailfoods . . . The Trouble with Nuts . . . Beware False Hunger . . . Adapting the Diet to Group Trips . . .
You Will Lose Weight

Would you alter your eating habits if you could double your energy on the trail while losing your love handles—without feeling hungry or deprived? I've discovered an eating program—not a diet—that does just that. You can lose excess weight and gain needed horsepower—by eating smarter, not less. Deanne and I have benefited greatly. Hundreds of thousands of others have enjoyed similar results, because the program is easy as well as healthy.

Since you presumably like to walk, you've got a big advantage over the couch potatoes who don't. Exercise is a vital part of the program. Frequent brisk walks will suffice. Daily twenty-minute aerobic walks will maximize results.

This revolutionary approach to nutrition has transformed the way I eat in the wilds—and at home—lifting me to a new level of lightness, power, and freedom on the trail. Adapting this program to wilderness travel brought me, almost overnight, a rebirth of youthful vigor and vitality—and unexpected weight loss.

It all began with a paperback I found in an abandoned car while hiking across a Hawaii lava field. The cover claimed it was "the No. 1 diet book of all time," so I took it home to Deanne, who'll read anything on the subject. She used to teach nutrition and cooking, and she's always gently nudging me toward better eating habits.

After a chapter or two, Deanne began reading me brief passages that excited her. At 5'10" and 148 pounds, I wasn't exactly fat. And I considered myself reasonably fit, with more than ample energy. Rarely a day passed without a brisk morning walk or half an hour on the basketball court before dinner . . . in addition to periodic tennis, swimming, skiing, and, of course, dayhiking. I'd never been on a diet in my life and I wasn't about to start now, but when Deanne got excited, I listened.

Big Breakfasts Are Bad

The first thing I liked about the diet was that it ridiculed the alleged benefits of the traditional all-American breakfast. All my life I've rebelled against big breakfasts on the grounds that it's dumb to eat when you're not hungry. When I ate I generally got heartburn, so I rarely ate breakfast, but I always felt a little guilty just the same.

The book offered a simple substitute for the obligatory morning pigout—eat fresh fruit or drink fresh fruit juice, nothing else. The authors claimed a fresh fruit diet would yield bountiful energy, a feeling of lightness, loss of excess weight, and healthful cleansing. And you could eat all you wanted. That was the second thing I liked. It wasn't a diet, the authors claimed, it was a lifestyle eating change. You didn't get hungry or feel deprived.

The third thing I liked was the program's commonsense basis. Experts, we're told, pretty much agree on what happens to food when it reaches your stomach and intestinal tract. They know how much energy it takes to digest each type of food, and which ones give back energy. And they know what happens when your stomach receives different food combinations at different times of the day. The basis of this program—oversimplified, of course—is to guide you to the foods that yield the most energy, and warn you away from the ones that steal it. It sounded sensible to me.

After a week of watching Deanne smack her lips over a big bowl of fresh fruit every morning and hearing her rave about how good it made her feel, I cautiously agreed to join her—since it wasn't a diet and she assured me I could eat all I wanted. But I warned her that fruit didn't always agree with me or even taste good. She assured me that was only because I'd combined it with other foods. Fruit, she informed me, must always be eaten alone.

I Come On Board

In the winter in Hawaii, we generally take a brisk thirty- to forty-minute, before-breakfast aerobic walk uphill from our village to get the juices flowing before it gets too hot. That's how we started the first day of our new diet—I mean program—walking uphill on empty stomachs. After we got back and showered I was hungry. Today, instead of toast and jam, cereal and milk and frozen orange juice, Deanne cut a fresh papaya in half, scooped out the seeds and filled the boats with sliced island bananas and grapes, squeezing lime juice on top. It was delicious, but I expected to be hungry in an hour. I wasn't. Or have heartburn. I didn't.

Instead, I felt light and bouncy all morning. Interesting! Usually after eating I feel heavy, lethargic, somewhat bloated. Today I didn't. I felt—light, as though I hadn't eaten, but I didn't get hungry that day until 11 A.M. Maybe Deanne was on to something. I decided to read the book and conduct a few experiments of my own.

The paperback bestseller I had found in the car was *Fit for Life II* by Harvey and Marilyn Diamond (Warner Books). By now Deanne had finished it and borrowed the original *Fit for Life* from the library. She insisted I read that first. While I read, I tested the "fresh fruit only until noon" ethic. I knew I had good energy until 9 or 10 A.M. while eating nothing at all. If I then ate fruit I found I got a tremendous burst of energy that seemed to last all day. If I didn't eat anything, I quickly got weaker and then hungry.

Experimenting with Fruit

For instance, one day I ate no breakfast and played tennis at 9:30 A.M. Feeling strong, I won the first set 6-2, but then I began to tire and lost the second 4-6. By lunchtime I was weak. I counted on a cheese and tuna sandwich to revive me, but it didn't. I felt heavy and uncomfortable all afternoon and I never regained my accustomed energy.

Another day, I ate fresh fruit after our walk and felt energetic all morning. I didn't feel hungry until 2 P.M., when I ate my tuna sandwich. This time it didn't upset my stomach. When I played tennis at 4 P.M., my energy was still so good that after losing 2-6 and 4-6 to a man twenty-five years younger, I won the third set 6-4 as he began to fade. Somehow the morning fruit had stayed with me. Clearly, there were giant implications for hiking in this program!

And after two weeks of nothing but fresh fruit in the morning, with no change in the rest of my diet, I stepped on the scales one day and discovered I had lost three pounds! Amazing!

Meanwhile, I read the book, learning why the morning fruit diet works. The familiar caveman image of man the meat eater, the authors claim, is deceptive. Long before learning to hunt, man subsisted on fruits and vegetables. A close look at our bodies yields overwhelming evidence that we were designed to eat fruits and vegetables, nuts and seeds, not meat, eggs, and dairy products. The nature of our teeth, jaw, stomach, saliva, digestive juices, alkalinity, and intestines seems to confirm this.

How Our Bodies Work

Experts agree, say the Diamonds, that our bodies can only run on one fuel: glucose manufactured by the body. Protein yields no glucose at all. None! Fat can be converted to glucose at the cost of great energy, but only if the body is starving. But fruit is full of fructose (fruit sugar) which is swiftly and effortlessly converted to the glucose our engines need. Use inferior fuel and you get poor performance. Eating foods that consume energy instead of producing it is like driving with the brake on. The laborious digestion of meat, eggs, and dairy products acts like a brake on the system.

According to the authors, fresh fruit (not cooked, canned, frozen, or dehydrated) zips through the stomach in twenty to forty minutes and into the intestines where it immediately begins generating energy. By comparison, any other food sits in the stomach for at least two hours—and more likely eight—before it's digested, consuming vast amounts of energy in the process. In fact the deliberate process of digesting most meals takes more energy than any other activity, including running and sex! During all this time and effort, no energy is being produced. My experiments seemed to confirm this.

After fruit in the morning, as long as I wait a mere half hour before exercising, I feel light and energetic. So I can swim or play tennis with energy and abandon less than an hour after my morning fresh fruit. In the old days, I always felt sluggish on the tennis court, even after eating two hours beforehand! Think what this means to the hiker getting ready for a strenuous day!

Eating & Exercising Strategies

Looking back, I see now that I could have rescued myself the day I played tennis on an empty stomach simply by eating some fresh fruit as soon as I came home, half an hour before any kind of lunch. The fruit would have turned to energy that fast. Because it's processed so quickly, I nowadays get a headstart on generating the day's energy by downing an orange twenty minutes before heading up the hill. I still beat the heat and I've found that my body will tolerate fresh (not acidic frozen) orange juice first thing in the morning. The quick passage of fruits through the stomach makes it easy to adapt this diet to busy schedules.

These experiments and others have caused me to radically revise my eating habits on the trail. The "fresh fruit only until noon" program is a perfect fit for the hiker who's already hooked on vigorous exercise—an essential ingredient for success.

The Rest of the Program

While fruit in the morning is the heart of the program, the rest is also applicable to dayhiking . . . though it's harder to implement. As always, the fundamental consideration is how the body reacts physiologically to individual foods—and their combination. *How* and *when* you eat are more important than *what* you eat or *how much.*

As we have seen, fruit alone on an empty stomach takes virtually no time or energy to pass through the stomach and begin generating energy. But fruits alone aren't enough for proper nutrition . . . or for most palates. By noon most people are hungry and want something more. Right behind fresh fruit in importance come the four "fruit vegetables": avocado, tomato, bell pepper, and cucumber. They are compatible with fruit, say the authors, and can be eaten for lunch without a waiting period. Since they double as veggies, they can be eaten anytime, while fruit can't. They take only about two hours to pass through the stomach before starting to generate fresh energy.

Next on the desirability scale come other raw vegetables suitable for the trail (carrots, celery, mushrooms, fennel, jicama, raw peas, broccoli, cauliflower, etc.). These are followed in ease of digestion by starches, grains, and breads. Last on the recommended list are raw (not roasted) nuts and seeds.

Hard-to-Digest Foods

To be scrupulously avoided in the Diamonds' scheme are meat, eggs, and most dairy products, because (1) they are believed to be completely nonessential to the body, (2) they steal energy from it for hours and often days, and (3) they fail to produce immediate fuel for the body. So there's no nutritional reason to eat them!

The persistent myth that we need to eat protein to build and maintain bone and muscle has been proven untrue, say the authors. My experiments and research have convinced me that I certainly don't need to eat protein while dayhiking, so I no longer carry meat of any kind on the trail. And the only dairy product I use is butter on my bread, because it's a fat, not a protein.

Combining different food groups, it has been proven, generally makes for digestive problems. Our bodies were designed to digest only one or two foods at a time. While fruit zips through an empty stomach when alone, if it's accompanied by any other food it rots uselessly, taking energy instead of giving it. That's why it only does its magic in the morning when the stomach is dependably empty. Once you switch to other foods, it's counterproductive to eat fruit again until digestion is complete (probably the following morning).

By the same token, when you switch to other foods, don't eat any more fruit. If you do, it will merely ferment instead of generating energy. Furthermore, you're likely to feel discomfort or heaviness and unfairly decide that the fruit didn't agree with you. The fruit stands alone.

Essential Liquids

Fresh fruit (and vegetable) juices are the ideal source of the liquid the body needs, being generally superior to water from any other source. Fresh juices provide the vitamins and minerals the body needs and can immediately utilize, while water commonly contains various indigestible minerals that can't be utilized. These are regarded by the body as toxic wastes and must be eliminated at the cost of vital energy. So, on dayhikes where water would normally be carried, it's smarter to carry peeled oranges for morning liquid nutrition, and fresh vegetable juice for afternoon.

Experts studying what happens to food after we eat it find our problems all stem from the way we combine different classes of foods. Alkaline foods require acid digestive juices, and vice versa. There are only two food groups, say the Diamonds, not four. So it's not surprising that combining opposite food groups leads to disaster as opposing digestive processes neutralize each other, insuring that nothing gets properly digested. When that happens, the beleaguered stomach simply gives up, dumping the whole rotting mess into the intestines.

If you eat fruits with proteins, for instance, the fruit ferments while the protein putrifies, both becoming useless toxic waste which requires the body's energy to eliminate. The same is true, unfortunately, of protein and starch (e.g., meat and potatoes). Two different proteins at the same meal (e.g., bacon and eggs) are even worse. The result of these miscombinations is poor digestion, weight retention, and a net loss of energy. Rotting food cannot be assimilated.

What's the significance of all this for the dayhiker? Our strenuous level of activity demands that our food *give* us energy, not steal it. Carrying rotting, energy-sapping toxic wastes in our stomachs as we head up the trail is like dragging an anchor. It's just as bad to carry food in our packs that will putrify and ferment as soon as we eat it. It therefore makes sense to eat intelligently before leaving and to only carry food that will generate, not consume, energy.

A Menu for Dayhiking

Preparation for a strenuous dayhike should ideally begin at dinner the night before. To insure complete digestion before morning, eat simply—and early. Go easy on fat, protein, eggs, chocolate, sugar, coffee, tea, soft drinks, milk, and ice cream. They're all extremely acid forming and/or difficult to digest, and they steal energy for long periods instead of producing it.

For best results, don't eat after 8 P.M. That's when your body shifts gears. Research, say the authors, shows that the body operates on three 8-hour cycles each day. From 8 P.M. to 4 A.M. the body is in its "assimilation" mode, absorbing and utilizing the food you ate earlier. So it doesn't want more food. From 4 A.M. until noon it's devoted to "elimination." That's why food requiring difficult digestion is unwelcome in the morning. From noon until 8 P.M. is the "appropriation" phase, when the body gratefully accepts food. Understanding these three phases helps explain the *Fit for Life* approach.

Smart Trip Planning

It makes sense to do most of your hard hiking in the morning . . . for several reasons. You're fresh and strong from a night's sleep, you're freshly fueled by the quick energy of fruit, and you're unencumbered by the substantial energy drain of digestion and fatigue. By afternoon, when you've switched from fruits to more complex and concentrated foods, a sizeable part of your body's energy will be devoted to digestion, not powering your lungs and legs.

We make a point of doing all our climbing, if possible, on fruit power, postponing lunch until we've scaled the peak or reached the pass. Sometimes we have to eat two or three fruit snacks on the way up, but we've found it well worthwhile to wait until the trail turns downward before switching from fruit to heavier foods. While climbing or hiking strenuously, we can stop briefly for fruit and keep going. We feel light afterwards and we soon get an energy boost. If we eat other kinds of food—except perhaps for candy, which gives a temporary lift but stresses the body—we feel heavy and are forced to slow down as digestion begins to compete with working muscles for energy.

Packing the Right Lunch

Next comes the choice of what to carry with you. For morning snacks (or second breakfasts) we start with a couple of peeled oranges or tangerines in a ziplock bag, a banana or two, and maybe an apple. Sweet seedless grapes are ideal, being full of liquid and perfectly packaged.

For backup (or a third breakfast) we carry various kinds of dried—not dehydrated—fruits, especially bananas, papaya, pineapple, raisins, and dates, all of it unsugared. Bone dry dehydrated fruits, we've found, don't provide us quick

energy. In the desert and in Hawaii, since we must carry all our water with us, we take extra oranges and a smaller canteen. Dates and bananas provide bulk and sweetness that are especially welcome later in the morning. We're also fond of dried apples, apricots, figs, peaches, and pears. Be sure your fruit is sun-dried, not sulphur-dried. It will cost your body valuable energy to eliminate toxic sulfur.

Once the switch is made to dried fruit, of course, it becomes necessary to drink lots of liquids for rehydration. Otherwise, dehydration can rapidly occur, with its threat of stomach weakness, cramps, dizziness, etc. So you need a substantial amount of water. But where should it come from? Should you seriously burden yourself by carrying it from home in a water bottle or canteen? Should you go to the considerable trouble of boiling it, filtering it, or adding chemicals? What to do about water has become a major problem for hikers in recent years, thanks largely to the Giardia scare.

Treating Suspect Drinking Water

As population and pollution increase, more and more of our once pristine water in the wilds has become suspect. The process has been accelerated in recent years by the threat of Giardia, a nasty though nonfatal protozoan infection that, we're told, is everywhere and has proved to be exceedingly hardy. Almost as mysterious as AIDS, Giardia has scared a majority of the people who've heard about it into taking elaborate steps to purify their drinking water. Giardia is no myth. A hiking partner of mine contracted it in Central America—

and suffered for a year. But I think there's been an overreaction. Fear of the violent symptoms—diarrhea, vomiting, abdominal cramps, fatigue, weight loss—has caused rational people to treat it like the plague. I regularly meet people in the wilderness who go to great lengths to purify their water—even after they learn that there's never been a documented case of Giardia in that region.

Let me therefore try to ease this blind fear by pointing out that Giardia has been around almost forever. It's only in recent years that the connection has been made between its symptoms and drinking unpurified water. In other words, if you've been drinking water in the wilds for forty years like I have, you've very likely been exposed to Giardia. If you haven't gotten it, it's probably because—like the vast majority of people—you're not susceptible to the little protozoa.

The U.S. Health Department estimates that somewhere between 1.5 and 20 percent of the U.S. population are carriers of Giardia, but are asymptomatic—because they apparently aren't susceptible. In that respect, it's a lot like the dreaded polio (infantile paralysis). More than 90 percent of the population has been exposed to polio and is immune. Only 1 to 2 percent of those who contract it suffer any degree of paralysis. So I have to question the need for treating ALL running water—just because it may contain Giardia protozoa. So does many a municipal reservoir!

Any authority you talk to will naturally take the most conservative position, to be on the safe side: "If in doubt," they'll say, "treat the water." Your family doctor, the Park Service, the Sierra Club, and the Forest Service all have to worry about being sued if they tell you it's safe to drink the water . . . and then you get Giardia. So do I.

But I continue to drink water I judge safe . . . without treatment, even when everyone else is treating it or lugging their own supply. I don't own or carry a water filter because I don't like the weight and bulk and the fiddling around. Filtering water is slow work and, some say, of questionable effectiveness. It's far from foolproof and may even be undependable. If I'm afraid to drink the water available along the trail, I much prefer to render it bacteriologically safe by adding a tasteless, iodine-based tablet to my quart (or liter) water bottle. Potable Aqua is the brand sold at REI. Just wait three minutes, shake, and serve.

The Sierra Club advocates adding two drops of fresh 7 percent iodine solution (a poison) to every quart or liter of suspect water. Shake well and let stand twenty minutes before drinking. But don't use it if you're pregnant or have thyroid problems. The U.S. Forest Service says, "Even clear running water should be purified before drinking. Boil it vigorously for at least three minutes before drinking. Chemical treatment or filtration may purify water, but is not considered as effective as boiling." Experts agree with Smokey. The Halazone tablets of yesteryear are no match for the bugs found in today's suspect water.

Today's hiker must take responsibility for judging the safeness of the drinking water supply along the way. I can't give you a rule of thumb, except to use common sense. Some people are so scared of Giardia and other pollutants that they won't drink anything that doesn't come out of a tap, carrying the considerable quantities they mean to drink. Others drink anything that halfway looks like water and rarely get sick. I regularly drink vigorously flowing water in the high Sierra as long as there aren't livestock or campgrounds upstream, but I realize I'm taking a chance.

A hiker should drink as often as he eats. If he is shirtless or the weather is warm, his body may easily lose a gallon of water a day! You can scarcely drink too much—provided you take only a little at a time. Severe dehydration will follow if lost fluids are not promptly replaced. In hot weather, make yourself drink before you're thirsty. By the time you're thirsty, dehydration has already begun.

Since water loss means salt loss, excessive sweating may justify taking salt tablets. A salt deficiency—from extreme water loss—can result in nausea, aches, or cramps. But overdosing on salt can be dangerous, too. Take no more than two salt tablets with every quart of water you drink.

After Fruit Comes Lunch

When fresh and dried fruits cease to satisfy, sometime around noon, we switch to the four "vegetable-fruits," tomatoes, cucumbers, bell peppers, and avocados. Since they combine well with bread as well as fruit, we often incorporate them into sandwiches. Just before leaving for the trailhead in the morning, we make sliced avocado and tomato sandwiches on buttered nine-grain bread. Sometimes we add lettuce and thin-sliced crunchy cucumber, bell pepper, or onion.

One of Deanne's greatest discoveries is Nu Tofu, a cheese alternative that comes in Jalapeño, Cheddar, Mozzarella, Herb & Spice, and Monterey Jack flavors. It adds real cheese flavor and welcome bulk to our sandwiches, without adding dairy (or lactose or cholesterol) to our diet. Deanne also makes an egg-free mayonnaise out of raw almonds, lemon, and safflower oil that's delicious. Finally, we add zest with mustard, Spike, and pickles. If we eat sandwiches for lunch, they're usually made, with variations, from the above ingredients.

Testing the Lunch Menu

The first time we followed this menu in Hawaii, we were descending a steep bluff that dropped 1,300 feet in two miles to Kealekekua Bay, to go snorkeling. I wondered if our diet packed the power to get us back up the hill in the 90-degree f heat of afternoon. After swimming for an hour with the tropical fish, we ate avocado/tomato sandwiches and crunched on raw fennel and corn chips. After resting for half an hour we were ready to climb out.

As we started up, I felt light and strong. Stopping only twice on the steeply climbing track to sip water, we made it back to the car, sweat pouring off us but still full of energy, in less than an hour. And I still felt light and strong. That trip convinced me. Since then I've been happy to rely on starch and vegetables for lunch—after a breakfast of pure fruit—no matter what the distance or terrain.

Never again will I burden my body with meat and cheese on the trail. For years I had thought it was propelling me up the trail. Now I see that it was holding me back. I was dragging an anchor in the form of energy-sapping digestion.

When we don't have the time or inclination to make sandwiches, we just grab an avocado and fill ziplock bags with cherry tomatoes, carrots, celery, and sliced bell pepper.

And for starch we eat hard crackers. Such a lunch can be thrown together in a couple of minutes, avoiding the preparation time of sandwiches. And lunch, augmented with raw nuts and seeds, can be munched all afternoon on the homeward trail.

To eat well and happily in the wilds, one must set aside the rigid and ritualized habits of urban eating—like the stricture to eat three square meals and not to snack between them. The opposite strategy is more appropriate on the trail, where the purpose of eating is to keep the body fueled and capable of sustained effort. Many snacks and small meals are easier to digest. Some successful hikers make it a rule not to let an hour pass without eating a little something. Others plan on eating two breakfasts and three lunches.

As *Freedom of the Hills* put it, "As soon as breakfast is completed the climber commences lunch, which he continues to eat as long as he's awake, stopping briefly for supper." In the morning, we start nibbling on fruit within an hour of breakfast, and we keep eating it as long as we're climbing. In the afternoon we "eat our way down the homeward trail," snacking on granola bars, raw almonds and leftover veggies—all brought along to keep the body well fueled.

"All Fruit" Days

Of course you don't have to switch from fruits to other foods at noon. *Fit for Life* recommends "All Fruit" Days for maximum energy and maximum weight loss. It's even easier for the dayhiker, who may well be off the trail near noon, to stick exclusively to fresh fruit while hiking—after an all fruit breakfast. You get all the benefits of energy and lightness while your body is working, and still eat what you like for late lunch and dinner.

If you're only going walking for two to three hours, you can probably eat most of your fruit before leaving, carrying little—if any—in your pack. And you still get all the benefits of the program. That's why it adapts so well to dayhiking. The biggest problem with an all-day walk is carrying enough juicy fruit to last all day. But the extra weight is worth it for the energy you generate and the energy you *don't* spend digesting.

Another option for the all-day hiker is switching back to energizing fruit after a well-combined (vegetable and starch) lunch. The necessary waiting period varies, depending on what you've eaten. According to the Diamonds, you can again eat fruit on a dependably empty stomach—two hours after nothing but raw vegetables, three hours after vegetables and starch, four hours after meat and vegetables (but no starch or cheese), and eight hours after eating without regard for proper food combinations.

Conduct Your Own Test

To test this eating program for yourself, I urge you to try this experiment. Pick a favorite strenuous dayhike and try it both ways. First walk it after a breakfast of bacon and eggs, toast and coffee, frozen fruit juice, oatmeal and the like, followed by a lunch of ham and cheese and mayonnaise sandwiches, fruit, roasted nuts, jerky, and all the trimmings. At the end of the trip, write down how you feel, your energy level, your tiredness, your heaviness, dullness, or fullness.

Then take the same trip, under the same conditions, but eating the new menu, starting with nothing but fresh fruit, switching around noon to fruit-vegetables, starches, nuts, and seeds. Eat no dairy products (except butter on your bread), no meat—and no fruit once you switch. At the end of the day make an honest comparison between how you feel now and your notes from the first trip when you ignored the way your foods were combined.

If you notice a decided improvement, take appropriate action when you plan the menu for your next hike. If you're as impressed as I was, you might consider altering your eating habits at home—especially if you'd like to lose some weight.

Evaluating Traditional Trailfoods

Now I want to comment on individual foods commonly carried on the trail in light of the new diet. I always used to take candy and sugar-rich foods—like instant oatmeal—for the "quick energy" I imagined they gave me. But refined sugar, I now believe, is a poison. It only gives the body a quick burst of unnatural chemical energy that doesn't last. It first stresses the body, then lets it droop. So now I leave the lemon drops home, along with the Wyler's lemonade mix, the Tang, and the jello. Chemical sugar substitutes are no better.

Granola and honey-oat bars are fine for afternoon, but I no longer carry concoctions that combine cereal and fruit. If you can't (or won't) restrict yourself to fruit, the next best strategy for breakfast is to eat only carbohydrates. I'd choose oatmeal and toast, with honey on either or both, with no milk, butter, fat, protein, or dairy to complicate digestion. For many years—until I discovered the power of fruit—hot oatmeal was my first choice if I faced a strenuous morning.

Chocolate combines sugar with caffeine, giving the body a wicked jolt and making the stomach scream. The climbers I know who eat it on the summit are lucky if it doesn't tie their stomachs in knots. Chocolate is a whip and ought to be reserved for dire emergencies. I never carry it.

Since I'm convinced that meat and dairy hold me back instead of helping me up the trail, I carry no meat of any kind (no tuna, deviled ham, corned beef, jerky, salami, steak, or chicken). And I've given up cheese, yogurt, margarine, eggs—everything except butter on my sandwiches.

I used to carry different kinds of gorp (basically roasted nuts and seeds mixed with chopped-up dried fruits) to "pick me up" in the afternoon. But of course it never did anything but fill me up, dehydrate me, and sentence my frustrated stomach to hard labor. Now I know why it never made me feel good. Its individual components are wholesome enough, but when they're combined they steal energy instead of giving it.

The Trouble with Nuts

When it comes to eating nuts, they ought to be raw, uncooked, and unsalted. That eliminates peanuts and all the commercially processed nuts and seeds we're used to. The easiest raw nuts to find are almonds and walnuts. Raw pumpkin and sunflower seeds are readily available. Nuts and seeds are much harder to digest than other sanctioned foods and should therefore be eaten sparingly and never in combination with anything but raw vegetables. We sometimes carry raw almonds for an afternoon snack, but we also break the faith by occasionally carrying roasted peanuts to eat on the downhill homeward trail.

Despite all these dietary restrictions, I never feel hungry or deprived. Instead, I feel emancipated from the energy-stealing foods that used to weigh me down and hold me back. I'll enjoy a tuna sandwich at home for lunch, but I no longer take one for lunch on the trail. From years of experience I know they're too heavy and hard to digest. If I'm going to be hiking after lunch, I want to feel strong and light.

You may well wonder what I mean when I speak of the "lightness" I feel with my new diet. Until I experienced it, I didn't realize I was used to a feeling of heaviness after eating all my life. I thought it was natural to feel lazy, bloated, and lethargic after meals. "Lightness" is the delightful opposite of the heaviness I always took for granted as a normal aftermath of eating. I never knew that the heaviness was unnecessary until it was gone. Try it and see for yourself.

Nowadays, when I think of bacon and eggs, meat and potatoes, or ham and cheese, I think of the unappealing heaviness they invariably brought. This new association helps me stay with a diet that keeps me eating and walking lightly. My aim these days is to make it up the mountain *because* of what I eat, not *in spite* of it!

Beware False Hunger

A related revelation involves the feeling of "hunger." It's easy to mistake lightness for hunger. And a feeling of hunger or emptiness invariably compels us to eat. We're conditioned from birth to eating until we "feel full," so it's easy to overeat in search of that familiar secure feeling—fullness. In the wilds the urge is intensified by an anxiety about keeping the body fully fueled. And it results in a powerful urge to overeat. It takes restraint and self-discipline to keep from gorging.

When I first started this diet I often felt hungry by mid afternoon. The hunger produced a powerful Pavlovian urge to eat. But every time I ate I experienced the old bloated feeling I have learned to dislike. So I experimented.

Finally I noticed that my mid-afternoon "hunger" was more like "emptiness" or lack of fullness. And it wasn't accompanied by any lack of energy. So I tried ignoring it, remembering that the Diamonds urge, "Don't overeat!" A slightly empty feeling, I remembered, often reflects a healthy lack of bulk in the stomach, not a dangerous shortage of nourishment. When I restrained myself from eating, I was pleased to discover that the emptiness soon vanished.

Now if I get the afternoon munchies I remind myself that my energy is good so my body doesn't really *need* more fuel. I'm just reacting to a lifelong habit—the urge to eat until I'm full. I turn my attention elsewhere and the next thing I know the emptiness is gone while the lightness and energy remain. If the hunger persists, as it occasionally does, then I know it's real and I eat nuts or buttered toast. Or if I'm relatively sure my meatless lunch has been digested, I happily reach for a banana.

A nice feature of this diet is its great flexibility. You don't have to deny yourself good food or suffer from real hunger or guilt. The dayhiker gets breakfast—the important part—out of the way before leaving, so he only needs to avoid injudicious food combinations on the trail. Once you get home you can eat as you please, but the better you eat, the more benefits you'll reap. Be assured that we don't follow the program faithfully at home. Far from it. We're often lured by the tug of habit, convenience, and the tyranny of our taste buds. And we often eat out.

On the trail, however, where we need all the energy we can get, we keep the faith. We go exclusively with fruit until lunch—or until we're through climbing—and we avoid meat, eggs, and cheese until we've taken off our boots.

My experiments confirm that this diet works. At least for us. After nine months, walking almost every day, Deanne and I both lost weight initially, then stabilized at low levels. And we haven't just imagined the increased energy and the appealing lightness. I've willingly changed my eating habits, and I've developed a new fondness for fruit. Since I'm careful not to combine it with other foods, it no longer bothers my stomach. Believe it or not, I don't even think of eating other foods for breakfast anymore. I look forward to succulent fresh fruit every morning. And I used to grumble at the thought of eating *any*thing before noon!

Adapting the Diet to Group Trips

Six months after converting to this foodstyle, we signed up for a ten-day Sierra Club Basecamp trip into the Sierra high country. I was resigned to sinking back to the traditional heavy camp breakfast, but Deanne convinced me that we needed our energy and lightness more than ever if we were going to be hiking at high altitudes every day. After several letters and phone calls to the leader, Deanne's persistence paid off. She was invited to bring a thirty-quart cooler full of whatever we wanted, and the cook would bend his menu to include a maximum of fresh fruits and vegetables in the eight coolers he was bringing.

Elated, Deanne bought a selection of green bananas, grapes, peaches, cherries, tomatoes, avocados, strawberries, oranges, apples, underripe honeydew and cantaloupe melons, crammed them in the cooler, and covered them with a bag of crushed ice. On our first morning at the trailhead she cut up the ripest fruits for us while the rest of the party dined on greasy sausage, scrambled eggs, oatmeal, and muffins, washed down with frozen orange juice and coffee. An hour later when we set forth up the trail, we bounded past the laboring group, our fruit already out of our stomachs and generating energy, while the rest of the party spent most of the day trying to digest its heavy breakfast—and walk at the same time.

By saving our dinner and lunchtime allotment of camp fruit for breakfast, to augment what we had brought in our cooler, we managed to eat not one but two bountiful fresh fruit breakfasts every single morning of the trip—with four apples left over. After eating a big bowlful of fruit with our companions every morning, we packed an apple, orange, and grapes—or the cook's raisins and dried apricots for a midmorning snack. And Deanne remained pure by skipping meat and cheese at lunch to dine on avocado, tomato, and lettuce sandwiches.

At first the group teased their two stubborn "fruitarians," but before long they made the connection: our morning energy and readiness to hike were undeniable. So was our lightness, slim fitness, and relative lack of thirst. While they lugged quarts of water or filtering apparatus, we drank sparingly from streams from a single cup. By the end of the trip, four of our new-found friends were eager to give the "morning fresh fruit only" scenario a try.

They had begun to see that the digestion of protein and fat demands the full attention of the body's resources for a considerable period of time. Even small amounts should not be eaten before or during strenuous exercise. The blood cannot be expected to simultaneously circulate rapidly through exercising muscles and digest complex food in the stomach without failing at one function or the other—usually both.

When menu planning, it will be helpful to match the fuel required by the body to the day's itinerary—in order to calculate how much food will be needed. Both the body's energy requirements and the energy production of food is measured in calories. For instance, it takes twice as many calories to walk at 3 mph as it does to stroll at 2 mph. Striding at 4 mph doubles the caloric requirement again. It takes two and a half times as many calories to gain a thousand feet of elevation as it does to walk at 2 mph. And it takes probably 2-3000 calories to propel a dayhiker all day.

You *Will* Lose Weight

Getting back to our early days on the diet, after just six weeks on fresh fruit only until noon, I found I'd lost eight pounds! Since then, my weight has stabilized at 140 pounds. I haven't been lighter since I got out of college. And I've never been so frisky or felt better in my life. Deanne, too, has lost weight effortlessly. She's tried dozens of diets but this is the first program that really worked for her. She no longer needs an annual fast to keep her weight under control.

Here's how I'd summarize my adaptation of the *Fit for Life* program for dayhikers. To get 70 percent of the benefits available, eat fresh fruit exclusively from the time you rise until lunch. For another 25 percent, skip meat, eggs, and dairy until you're off the trail. The final 5 percent comes from eating sensibly the night before.

The choice is yours. You can trudge up the trail with a miscombined breakfast rotting in your stomach, expending vast amounts of energy to digest food that will never produce energy—essentially hiking with the brake on. Or you can charge yourself with fruit, doubling your energy, freeing your muscles to propel you lightly up the trail in high gear, happily walking off excess weight as you go.

If you think this eating strategy is radical—as I would have five years ago—face the possibility that you're behind the times, out of step with modern research.

There is ample evidence that the traditional All-American diet is a disaster. Nearly two thirds of the nation is at least twenty pounds overweight. Some 200,000 bypass operations are performed each year because our arteries are clogged. The best-selling prescription drug in America today is an anti-acid. Heart disease kills four thousand Americans every day. "A vegetarian diet," says the conservative AMA, "could prevent 90 to 97 percent of heart disease deaths."

Clearly, meat is the culprit. Maybe it's time to consider the alternative, since meat isn't essential to nutrition and doesn't provide any great benefits. If, like me, you aren't quite ready to go vegetarian, the *Fit for Life* diet is a good place to start—especially if you'd like to gain energy and lose weight.

SEVEN

Extended Dayhikes & Climbs

Finding Dream Trailheads . . . Roadhead Rustic Lodgings . . . Wilderness Camps . . . New Zealand Hut Hiking . . . Hut to Hut in the Alps . . . Shuttle Dayhikes . . . Yosemite Backcountry . . . Raft & Ski Trips . . . Hiking Club Outings . . . Yosemite's High-Country Camps

The ultimate dayhike, it seems to me, is one that doesn't end at sundown. Magically it goes on into the night and the following day, even day after day. Somehow you live in the wilds without a pack. Like a bird, you're free to wander the wilderness unburdened. At first that seems to defy the very definition of dayhiking, but I described one way to manage such a feat at the end of chapter 2. By choosing a trailhead that offers quality car camping and a variety of trails, we surmount dayhiking's one drawback: coming home at night.

There are an almost unlimited number of trailheads that qualify. You only have to know what you're looking for to find them. And there are many other ingenious ways to keep the adventure alive without carrying a backpack. I'll discuss them one by one, starting with places you can drive to that are free, moving to commercial establishments on roads, following with schemes and outfits that lie beyond the roads, and ending with the offerings of hiking clubs. Specific examples of each are included.

Finding Dream Trailheads

The best places to find dream trailheads are in national parks, around the edges of wilderness areas, in national forests, state parks, and other scenic areas. These government recreation areas are most likely to provide both multiple trailheads and attractive campgrounds that are free or inexpensive, and it's usually easy to get needed information because park personnel are hired to help the public by answering questions, dispensing maps, and selling books and guides.

In addition, in the West at least, by studying maps, making local inquiries, and driving around, it's often possible to find lonely road ends where it's permissible to camp. Then there are logging roads, jeep roads, mine and mill roads, ranch roads, abandoned roads, open range, and so forth, where you can find your way to wilderness or sometimes camp within walking distance. Even if camping is prohibited, you may be able to dayhike. By nosing around, especially in the western deserts, there's a good chance of finding an exceptional opportunity.

Roadhead Rustic Lodgings

Next come commercial lodgings at appealing trailheads. All along the margins of California's Sierra are roads winding up canyons that end at trailheads, usually on lakes or streams. Here can be found all manner of quaint and often inexpensive overnight accommodations: fishing camps, inns, housekeeping cabins, lodges, campgrounds, bed and breakfasts, tent cabins, and rustic motels, often with tiny general stores, cafes, or dining rooms. There may be boats and horses for rent, guides and packers for hire.

I find these rustic lodgings so delightful in themselves that the adventure does not end when I return from each day's hike in the wilds. There are charming inns that provide me a room with the sound of a trout stream and a view of a lake or peaks that also offer trails to intriguing nearby wilderness. I am happier spending a long weekend in such a place, dayhiking every morning after breakfast, than I would be on the majority of backpacking trips.

Others have discovered the same allure—and packaged it. Palisade School of Mountaineering runs a wilderness camp eight steep miles from a trailhead near 14,000-foot Mt. Whitney, where fledgling mountaineers can go to learn their craft. But that belongs in another category. Palisade also offers three-day guided dayhiking trips it calls "Skywalks," making use of rustic inns and lodges. For about $200 they take you to four of them. You drive to the first, stay the night, and spend the following day hiking to the next one, carrying only a daypack. After three dayhikes and four lodges, they shuttle you back to your car.

Wilderness Camps

Some of the best base camps for dayhiking lie beyond the most primitive public roads. Accommodations usually cost more than at trailhead inns, but by providing an escape from the sight, sound, and smell of cars, they offer more of the feel of wilderness. Finding these widely scattered hostels is the problem. There isn't any directory, and most don't advertise, but I can tell you where to look. Many of them are ranches that take guests, so query packers and ranchers associations. Ask park information people, chambers of commerce, and inquire at all those rustic inns at trailheads. Sometimes you'll find classified ads for these lodges, camps, and ranches in the back of outdoor magazines. The wilderness travel companies are another likely source. Finally, go hunting where you *want* to find them or suspect they exist. Like lost arrowheads, they're waiting to be found. Try the buffer zones around parks and wilderness areas.

A second possibility is the strings of summer tent camps, huts, or cabins that lie beyond the end of the road. Whereas the ranches and base camps described above are single destinations from which hiking trails may radiate, these strings of accommodations are linked together by a trail and set up for the hiker who wants to move through wild roadless country without carrying a pack. Usually they provide lodgings and meals. Since these camps lie beyond the roads, they offer all the benefits of camping without the drawbacks, unless true solitude is your aim.

For me, in terms of lasting memories, the biggest feature of these two types of trips is the people I meet, the new friends I make, the close companionship that so often evolves when strangers are thrown together in the wilds. When I want solitude, I find it during the day in solitary dayhiking. In the evening, it's delightful to sit around the campfire and visit with like-minded outdoorspeople in a lovely setting. That's why I consider trips of these two types as the ultimate in dayhiking. Consider a few examples.

New Zealand Hut Hiking

The Southern Alps of New Zealand offer two different types of hut accommodations: public and private. In rain forest where precipitation reaches 240 inches a year (that's twenty inches a month!), huts are more practical than tents for staying dry. Deanne and I recently returned from a guided, private, five-day, four-night tramp called the Routeburn Walk, which offers more time above timberline than its more famous (but crowded and expensive) neighbor, the Milford Walk. We made reservations for this trip in the United States from brochures obtained from the New Zealand government's tourist information agency.

At the trailhead, after a bus ride around Lake Wakatipu from Queenstown, our local guides issued us serious rain gear, sleeping bag liners, and lunches to stick in our daypacks. Before we'd been on the trail an hour it was raining, and by the time our international party of eight reached the hut, we were soaking wet. After hanging our wet clothes, boots, and socks in a marvelously efficient drying room, we picked out bunks with mattresses in the snug dormitory and inserted our liners in the waiting sleeping bags. Then we quaffed hot tea by the fire, took hot showers, and got better acquainted with our companions from five countries while our guides prepared dinner.

Meanwhile, less than a mile away, the backpackers (called "freedom walkers" down under) who had traveled with us on the bus were drying their clothes over a stove in the less elegant public hut, unpacking their sleeping bags and food, and preparing to take turns cooking their dinners in the community kitchen. In these huts, a resident manager maintains minimum decorum and sees to it that hygiene is reasonable and the hut is swept clean before the trampers depart the next morning. Water is drawn from a nearby stream and there are no showers, but the huts are free, although reservations are required.

Each day we met the heavily laden freedom walkers along the trail. Sometimes we all ate lunch together on the heights, and we visited back and forth in the huts at night, trading experiences in the lovely New Zealand high country. This dual system of huts, though not everywhere available in New Zealand, provides an ideal example of two ways to travel the same wilds, with differing costs and comforts. Our guided dayhike on the Routeburn cost about $400, compared to more than $700 for the Milford Walk.

Some years before, I took my family on another variation of hut-to-hut travel in New Zealand, alternately traveling on foot and by jet boat. Adjacent to the Routeburn Walk is the Hollyford Valley, famous for its trout fishing. The first day, we hiked eight miles down the river to a private hut (provisioned by helicopter) with accommodations similar to those we enjoyed on the Routeburn. On the second day, a jet boat took us hurtling down the raging Hollyford River, bouncing off sandbars and stopping in quiet water to show us schools of two-foot trout. For six days we traveled along the river in this manner through Mount Aspiring National Park, from the interior to the ruins of a gold rush town at the sea.

Hut to Hut in the Alps

I wish I could say that America boasts many similar opportunities for hut-to-hut travel, but it doesn't. We lag far behind other countries. There are extensive hut systems in Europe that permit everyone from casual strollers to skiers and serious mountaineers to dayhike in comfort. Throughout Austria, Switzerland, eastern France, and northern Italy there are strings of huts, chalets, inns, and pensiones that permit dayhikers to walk for weeks without staying in the same place twice. Some of these huts are more like hotels, with waiters in tuxedos serving cocktails on the terrace. Others are spartan but clean and warm.

An Italian hiking companion of ours in Hawaii named Andy travels halfway around the world every few years to spend a few weeks hiking from hut to hut in the Italian Dolomites, part of the Alpine massif that stretches from France to Yugoslavia. A rental car takes him to a mountain village, where he makes his hut reservations at the town tourist center. Then he rides a tram to the first of a string of timberline huts at 8,000- to 11,000-foot elevation. For days or weeks he then hikes the Italian high country, finding comfortable lodgings only two to three hours apart.

Lodging in a private room or dormitory costs him as little as $10 a night. Meals, featuring meaty soups and spaghetti, are more expensive. Andy especially enjoys the stimulating companionship of the outdoor people from around the world that he meets in the dining room and on the trail. There are hundreds of huts and thousands of miles of trail in the Alps. The network is so vast that a hiker could sample them all summer for years without repeating any. When Andy has completed his circuit or is ready to come down, he simply gets back on the tram and rides down to the village and his waiting car. It sounds like a very civilized way to walk the wilds. One of these summers we're going to go with him.

Besides these posh minihotels, there are unstaffed yet comfortable huts perched high beneath the famous peaks, waiting for use by serious climbers. I'll never forget the one I visited just after my twenty-first birthday. Three nights before, in a tavern in Zermatt, Switzerland, I found myself drinking beer with an Englishman destined to become one of Britain's most famous daredevil climbers, Ian Gordon McNaught-Davis.

We found we shared the same birthday. He said we should celebrate by going climbing. I must have exaggerated my mountaineering experience, because the next day he informed me we were going to try a new route on an "easy" mountain. I hadn't any equipment, but Ian quickly produced everything we needed: boots, ropes, ice axes, packs, sleeping bags, climbing hardware, food. Carrying awkward rucksacks, we set off from Zermatt for the climbers' hut close beneath the peak. It perched on the moraine near the head of the Rhone Glacier at close to 13,000 feet. It was empty, but there was wood for the stove, water, and a wall-to-wall bench with side-by-side mattresses that would sleep a dozen climbers. Ian made dinner and promptly went to bed, instructing me to do likewise.

It was 4:00 A.M. and dark when he woke me and gave me a bowl of hot oatmeal. Minutes later we were roped together and climbing the glacier, crossing blue crevasses on ice bridges in the growing light of dawn. Hours later, we left the glacier, moving up a steep knife-edge of ice, Ian walking nonchalantly while I crawled. By midmorning we reached the summit, and I was shocked to find myself looking down on the Matterhorn!

This "easy" mountain was the Dufourspitze of the Monte Rosa, the highest mountain in Switzerland! But there wasn't much time to exult. Now came the "new route." Ian wanted to drop straight down the cliffs to the north to the glacier below, because it had never been done.

I'll never be sure how we made it down, belaying one another, often swinging on pendulums over overhangs, then shaking the rope loose from the snowy rocks above. At dusk we reached the glacier and staggered back down through the crevasses to the hut. I vividly remember falling repeatedly on my face from exhaustion. When we reached the hut, neither of us could climb its three steps to the door. We crawled. And once we got inside, we were unable to rise. We crawled across the floor to that community bed, climbed up on it, and collapsed. It was hours before we laboriously untied the rope that had joined us for fourteen hours. Then, lying on our backs in the dark, we talked half the night before we were calm enough to sleep.

By noon the next day we had sufficiently recovered to leave the hut that had saved our lives and made our climb possible, and hike down to Zermatt before dark, planning as we walked to climb the Matterhorn next, then move up the next year to the Himalayas to try for some first ascents. A week of snowstorms prevented us from attempting the Matterhorn, and a lack of cash for porters blocked our trip to Nepal. But years later, Ian made it to the Himalayas with the British Baltoro expedition for a number of noteworthy first ascents.

Shuttle Dayhikes

Although hut systems in America are comparatively rare, there are other types of accommodations that permit the summer dayhiker to remain in the wilderness without a backpack. Thirty years ago, I was fortunate enough to buy a ramshackle old cabin on the edge of Desolation Wilderness, several miles beyond the road and accessible only by boat or trail. It took me six summers to fix it up and make it liveable, but it's been a great base camp for dayhiking ever since. (My experiences from the first fifteen years there are collected in my book *Mountain Cabin,* also published by Ten Speed Press.)

After backpacking and dayhiking all of Desolation's trails for years and writing a guidebook to the area, I began to search out the hidden jewels between the trails and climb the peaks, dayhiking cross-country. Technically, after all, wilderness doesn't truly begin until the trails are left behind. To find fresh country to explore I had to get more inventive.

For instance, an old map showed a gold rush dairy in a distant area I hadn't visited, with no nearby trail. A backpacking expedition wasn't feasible because a cliff stood in the way, so I plotted a long one-way dayhike that would leave from the cabin, traverse the wilderness, and end at a trailhead on the far side. After arranging for my wife to pick us up there at dusk, my friend Galen and I started off. At the end of the trail we moved into new country. Wearing only daypacks, we found a safe route up the cliff.

By late afternoon we had found the old dairy, now a high-country cow camp. We also found the rancher whose family had owned it for four generations. He showed us the 150-year-old buildings and told us hair-raising tales of wildcats and wolverines, rustlers and bandits. It was nearly dark when we made it out to my patient wife and the waiting car. A shuttle had enabled us to explore new country and reach our objective on a memorable dayhike. We couldn't have done it with backpacks.

Yosemite Backcountry

Private mountain cabins in the wilderness are exceedingly rare, of course, but there *are* public facilities so delightful and intriguing that we leave our cabin every summer to visit them. We love the variety and companionship, the fresh faces and new country. The high country of Yosemite National Park is a second home to me. I first discovered its string of summer tent camps joined by a trail when I stumbled onto Vogelsang High Sierra Camp at 10,200 feet on a backpacking trip while still in high school. By buying a few staples there, my companions and I were able to extend our trip another week, living mostly on trout—until my mother sent the rangers out to find us.

My first wife and I spent our honeymoon at Vogelsang, and Deanne and I leave our cabin to go there for a few days every summer. The string of six backcountry camps forms a rough circle that runs through Tuolomne Meadows on the Tioga Pass Road. Tent cabins are clustered around bathrooms with hot showers and flush toilets. There is family-style seating in a community dining tent where excellent breakfasts and dinners are served. Lunches may be ordered for the trail. The camps range in elevation from 6,500 to 10,200 feet and are located a day's walk apart (six to ten miles).

Raft & Ski Trips

Raft, kayak, and canoe trips offer another way to travel through wilderness without carrying a pack. I was first drawn to rafting by the thrill of running whitewater. Along the way I came to realize that I could also make long traverses through beautiful wild country, enjoying all the comforts of car camping, without carrying an ounce on my back. And I could sleep in true wilderness by the sound of running water. To fully enjoy wild rivers in a raft, I wanted the fun of navigating them myself, so I trained to become a river guide. (See my *Whitewater Boatman: The Making of a River Guide,* also from Ten Speed Press.)

Whitewater rafting, though often thrilling and sometimes frightening, is surprisingly safe—much safer than driving America's highways. The majority of trips, like dayhikes, are over before dark. You make a reservation with an outfitter, show up early in the morning at the "put-in," divide into crews of six paddlers, and learn to paddle together from your captain/guide in gently flowing water. Then you're off down the river. You stop somewhere and go ashore to eat the lunch provided by the outfitter on a sandbar. Then in late afternoon you help pull the rafts from the water at the "takeout." When they're loaded for the return trip, a bus takes your happily weary crew of new friends back to your car at the put-in. Such day trips usually cost $60 to $120.

Overnight trips, though they require the planning of a camping trip, supply all of the advantages of both dayhiking *and* backpacking, with none of the drawbacks. We love to run the desert rivers because the trips are longer. On multiday trips, one quickly gets into the rhythm of life on the river, pulling in to shore to make camp on a sandbar in the afternoon, helping with the chores, sitting by a campfire in the dark after dinner, listening to the whisper of the river or the boom of the rapids to be run the next morning.

Sometimes there are layover days or short days to permit a longer dayhike up a tributary canyon. On most days there is time before dinner for a walk along the canyon or the climb of a butte for a taste of solitude and quiet. We like the five- and six-day trips offered on the Owyhee River in eastern

Oregon; the Yampa, Green, and upper Colorado in Utah; the Dolores in Colorado; and the Salmon and Snake in Idaho. The ultimate trip is the twelve- to fourteen-day journey through time down the 225 miles of the Grand Canyon in Arizona. Outfitters customarily charge $75 to $120 a day for these longer river trips.

On some rivers it's feasible to travel on your own in kayaks or canoes instead of rafts. You're responsible for getting your craft down the river, while the outfitter/guide navigates the supply boat. Often trips are combinations. Because I've been a guide, I usually negotiate with the outfitter to captain my own raft, usually a gear boat. On a five-day run down the Dolores in Colorado, for instance, there were paddle rafts with guide/captains, our gear boat, kayakers (both hard slalom boats and clumsy inflatables), and a couple of Eastern gentlemen in a whitewater canoe. We all came together at night in camp, and often we traded boats during the day.

Another enjoyable form of daytripping without a pack emerges when snow covers the land, hiding the trails (and more importantly the roads). A fresh covering of snow is more than a blanket of frozen water. It's a medium that offers swift travel in a new world of white. It opens access to country that otherwise could not be reached in a single day. When conditions are right, seasoned cross-country skiers can travel twenty miles over terrain that would limit them to ten without snow. And there's a freedom that has to be experienced to be comprehended.

Deanne and I can cross-country ski to our cabin in the winter across frozen lakes, without regard for roads or trails, in half the time it takes us to walk to it in the summer. We travel light, carrying fannypacks or daypacks. And in the spring we often strip to swimsuits or less as we ski across the ice in temperatures that reach the eighties, our clothes heaped on our packs. After stopping at our cabin for lunch, we can zip across the upper lake and into the wilderness, returning downhill to the cabin by sundown.

Snowshoers have less range, but they can also enjoy the freedom of traveling cross-country on terrain that can't be managed without a covering of snow. For instance, since we're not expert skiers, we climb the steep mountain behind the cabin on snowshoes, secure in our ability to navigate steep icy slopes without crashing. And when conditions favor snowshoeing over skinny skiing, we are happy to switch to an easier way to travel the winter wilds. Both cross-country skis and snowshoes can be rented until you're ready to make a modest investment.

Hiking Club Outings

One of the best ways I know to extend daytrips and enjoy the companionship of like-minded people is to travel on guided trips with hiking clubs. I've been a member of the Sierra Club for nearly forty years, not only to support its work in conservation, but also to travel with leaders who generously share their knowledge of choice (and often hidden) places in the wilds with fellow members and the interested public. It's an unbeatable way to get acquainted with new country and new people.

When Deanne and I moved over the mountains to the western side of the Sierra, we shifted within the Sierra Club to the Mother Lode chapter. Now we get its monthly newsletter, which lists myriad chapter outings, often three or four a week. But we retain our subscription to the Great Basin chapter's newsletter to keep track of upcoming intriguing adventures. Trips are rated according to difficulty and needed experience.

In addition to the free local outdoor programs of each chapter, there are National Sierra Club outings of every kind all over America and in many foreign countries. As with commercial trips, costs run $50 to $150 a day, plus transportation to the trailhead. A great many of these outings offer superb dayhiking opportunities. Our favorite types are "Highlight" and "Basecamp" because both permit us to enjoy all the benefits of dayhiking without the drawback of ending the trip at dark. We have also taken National Service and River trips.

Every January we pounce on the annual outings issue of the club's national magazine, *Sierra.* We eagerly devour the brief listings of hundreds of trips worldwide in search of outings that excite us. Then we order detailed trip accounts from the Sierra Club office. Highlight and Basecamp dayhiking trips have a lot in common. On both, provision has been made to transport twenty-five pounds of each tripper's personal gear (usually on pack stock) in addition to all food, equipment, and community gear. You carry only your lunch, camera, rain gear, and so forth in a daypack. Trips usually last between one and two weeks.

On Highlight Trips, the party has an itinerary and keeps moving, usually traveling in a rough circle or employing a shuttle to avoid retracing its steps. Camps are typically three to six dayhiking hours apart. After several traveling days there is usually a layover day. All participants must be club members. No guns, pets, or radios are permitted. Children must be at least nine years old and accompanied by a parent. Otherwise there are no restrictions. Trippers help set up camp, assist the cook (provided by the club), wash dishes, and gather firewood. Participants are diverse, but all share an appreciation of the wilds and a wish to preserve it.

Basecamp Trips are similar, but instead of packing up and moving to a new campsite every few days, the group stays in a single camp, anywhere from two to twelve miles from the trailhead. Participants dayhike in, set up their tents and the kitchen, and bid the packer good-bye. The camp then becomes home base for dayhikes (and perhaps backpacks) into the surrounding country for the duration of the trip. You're free to do whatever you want—from loafing around camp to making strenuous climbs—only assembling for breakfast, dinner, and chores.

Both Basecamp and Highlight trips are a dayhiker's dream: living in the wilds and exploring the country day after day without the need to carry a backpack. Highlight Trips cost more because the packer and his stock stay with the group. They are also more strenuous, usually covering fifty miles or more of trail, with perhaps some cross-country travel as well. Basecamp Trips, being less structured, offer more freedom, with far more time for fishing, swimming, photography, reading, bird watching, climbing, contemplating. You set up the kitchen and your personal camp only once.

By chance, Deanne and I made trips of each kind, two years apart, into the High Sierra granite country of Ansel Adams Wilderness, just south of Yosemite National Park. Lying well west of the John Muir Trail, this wilderness area is lightly traveled, but it boasts a magnificent view to the east of Banner Peak and Mount Ritter and the sawtooth skyline of the Minarets. Our Highlight Trip lasted seven days and cost us $620 each. Starting at 7,000 feet, carrying only daypacks, we set forth in mid-September through open forest, flower-strewn meadow, and glacially smoothed granite, climbing toward Yosemite.

Most of our twenty companions carried full-sized backpacks because they couldn't make the strictly enforced weight limit for the pack stock—twenty pounds per person on that trip. For us, carrying backpacks would have defeated the trip's main purpose. On the fourth day, we crossed 11,000-foot Isberg Pass into the park. On a layover day, four of us climbed 12,000-foot Foerster Peak. Then we descended from the park to the San Joaquin River and circled back to our original trailhead without retracing our steps. Deanne and I made several lasting friendships among the twenty-three trip participants from four states, England, and Canada. Most were singles and couples in their thirties and forties. Despite traveling two days out of three, there was ample time to catch enough trout for all who wanted them. There were no bears, mosquitos, or threats of Giardia.

On the Basecamp Trip two years later, in early July, twenty-three of us congregated at a nearby trailhead to meet our leaders. We were issued tiny squeeze bottles of 7 percent iodine to treat drinking water—if we wished. Two drops in a quart or liter, we were told, would render it safe after a vigorous shaking and a twenty-minute wait. This was an older group, probably averaging fifty years old and running from sixteen to seventy. Our duffel was weighed, and packs of more than twenty-five pounds were returned to their owners for lightening. The same packer as before hauled duffel and provisions for nine days to Vandeburg Lake, just below timberline at 8,800-foot elevation.

We had all day to make the five-mile, 1,200-foot climb to camp. As before, Deanne and I carried only daypacks, while our companions without exception trudged beneath variously loaded backpacks. Powerfully fueled by fresh fruit, we quickly left the rest of the laboring group behind, taking a cross-country route to see more wilderness and wildlife, arriving in camp before the others, still fresh.

Once the kitchen/commissary was set up and our provisions safely stored in the cooktent, we pitched our tents. Then we were free for nine glorious days to dayhike in every direction, visiting remote lake basins for trout, climbing peaks, swimming, getting acquainted with our fellow campers. Four lively couples of veteran campers who had made many trips together invited us to join them, and we became good friends before the trip was over. The necessity of battling early season mosquitos and almost daily thundershowers united the whole group into a tightly knit unit.

On the final day, we walked back to our waiting cars at the trailhead while the packer hauled everything back to the pack station. The ten-day excursion had cost us $495 each. Fully half of that went to the coffers of the Sierra Club to support its outing program and other good works. After considering the trip food budget ($1,700) and the packer's spot pack charges, we and our eight new friends decided to organize our own spot pack base-camp trip for next summer.

At the pack station we learned that the charge for a packer and horse was $70 a day, and a mule able to carry 150 pounds cost $40 a day. Even after hiring the same cook, we could cut the trip cost in half, and have everything exactly the way we wanted it, without a leader to tell us what to do. It means a lot of planning and scouting and organizational work, but our group possesses the necessary experience and compatibility to make the trip happen—and succeed.

Thus, privately organized spot packing offers yet another way that dayhikers can escape the confines of their sport and extend the wilderness experience without carrying everything on their backs. And it's economical for even one or two people—if they're willing to do their own cooking. Somewhat more expensive are "packer accompanied" private trips, the format for our earlier Highlight Trip.

Before moving from volunteer nonprofit trips run by clubs to commercial, profit-oriented trips, I have to warn that

Sierra Club trips are not for everyone. While I've been a Sierra Clubber for forty years and heartily support the vast majority of its activities, I feel the club's national outing program has suffered a mild decline in recent years, due partly to its rapid growth and the rising cost of insurance. A decade ago, its offerings were superior to those of most commercial outfitters in price, leadership, and clientele. Now its prices are often higher than those of commercial outfitters, its water trips have all but dried up, and leader quality has slipped.

With more than eight hundred volunteer leaders, it's inevitable that some will be insensitive or incompetent, prone to getting lost or more in tune with nature than with the needs of their charges. The national program now has nearly three hundred trips annually (with thousands more at the local chapter level). There is something for nearly everyone.

A major dividend of group trips is the chance to form friendships with like-minded people. Deanne and I go on group adventures for the people we meet. We are planning a spot pack trip with the four couples we met on the Basecamp Trip. And we've invited them to accompany us on one of several trips to Alaska. We'll either do a float trip down the Copper River, a tributary of the Yukon, and dayhike off the river, or do a combination dayhike–river trip on the north slope tundra of the Brooks Range.

Both trips are with Nichols Expeditions, a tiny but wide-ranging and inexpensive outfitter consisting of friendly, capable Chuck and Judy Nichols. We ran the Snake and Salmon rivers with Chuck and Judy a few years ago, when they willingly provided me with an oar boat of my own. And we celebrated the completion of this book by joining them for a week of sea kayaking in Mexico's Sea of Cortez. They also do hill walks in Thailand, mountain biking in the Utah deserts (from their bed and breakfast in Moab, Utah), and trekking in Peru and Patagonia.

Speaking of effective, reasonable, personable outfitters, yet another dividend of our Sierra Club Basecamp Trip was an introduction to Cathy Harrison, the bubbly Dutch proprietress of Knapsack Tours. Cathy's approach to extended dayhiking is a little different. She takes her groups to a region offering a wide variety of dayhiking trails of differing difficulty. She houses the group in small, no-frills, yet comfortable and friendly lodgings. Then, for two to three weeks, she whisks them by van every morning to the appropriate trailhead in small groups for the day's guided hike. There's a short one for those who only want to stroll and smell the flowers, and another for walkers who want to see a lot of country.

Her efficiency, warmth, and budget prices have built her a large following. While most commercial, guided group hikes cost the client more than $100 a day, plus transportation to the trailhead, Cathy's Knapsack Tours in 1990 (nicknamed "Shoestring Tours") averaged $70 a day for three weeks in the huts and hotels of the Swiss Alps, $90 a day for a week in the inns of Vermont to view the fall colors, $60 a day for two September weeks in the lodges of the Canadian Rockies, and less than $50 a day for five to six summer days of dayhikes in the Yosemite backcountry.

Yosemite's High-Country Camps

What could be more fun than wandering the glaciated granite high country of Yosemite for a week, wearing only a daypack? Mentioned earlier, the six tent camps are spaced a day's hike apart on the justly famous Yosemite "loop." We began at Tuolomne Meadows at 8,600 feet, the only camp reachable by road. Packing for the fifty-mile loop trip was logistically much different than preparing for either the Basecamp or Highlight trip. There would be no stock to carry our camping gear, because we didn't need any. We would stay in white canvas tents furnished with beds and a potbelly wood stove. We would eat family style in a canvas dining room and enjoy hot showers and flush toilets, but we had to carry everything we might want or need with us—clothing for all contingencies, toiletries, camera and fishing gear, reading matter, first aid kit, spirits, and so forth—for seven days.

Deanne and I amassed a long list of necessities and started whittling it down. We couldn't whittle much clothing because snow was forecast for the High Sierra for the first five days of our journey. Failure to stay warm and dry could easily ruin the trip. It's miserable to be miserable—even in Yosemite. At the same time, we were determined to go light, to limit ourselves to daypacks. To load ourselves down with backpacks would have defeated our purpose. We wanted to be free to play along the way, to be fresh for side trips away from the trail, to exercise the trout and climb a few peaks.

Because of the forecast, I left home my big Lowe fannypack. With its drop-seat extension it might have handled the bulk, but the weight would have been excessive. I dislike a pack on my back, but more than six pounds in a fannypack makes it bounce uncomfortably. So I decided to carry my twenty-year-old North Face daypack, while Deanne took the bigger Lowe climber's pack I had bought her for the occasion, partly because she planned to carry more clothing. I decided to depend exclusively on vapor barrier warmth, leaving behind all insulated garments. To even the loads we decided to switch packs periodically.

We hoped to somehow continue our fresh-fruit-only-until-noon diet, but we were given no assurances about the fruit supply available in the camps. So we carried a supply of our own air-dried Hawaiian bananas, raisins, dates, and dried apricots, plus three avocados for lunch.

After two days of acclimatization on cross-country tune-up walks and scouting trips out of Tuolomne, we set forth. I carried about twelve pounds, Deanne about eighteen, as we followed the river downstream. When we attempted to trade packs, we found hers was too small for me. After that, to even the loads, I carried the food and other heavy items while she took the bulky clothing. The first night in camp was bitterly cold and windy. If the racing clouds had ever slowed down, it would have snowed.

After a hot shower and a few drops from my flask of 151-proof rum, we went to dinner, where I wore vapor barrier socks (plastic bags) beneath my socks in my boots. Over my polypro long-sleeved shirt I wore a VB shirt, then a hooded sweatshirt, topped by a nylon hooded shell. On my hands were mittens over plastic VB painter's gloves. On my head was a watch cap, hood, and shell. And I was warmer than some of those who shivered in bulky padded parkas.

After dinner, I peeled it all off, layer by layer, by a roaring fire in the potbelly stove in our toasty tent, which we shared with another couple. In the frosty dawn, before the sun reached our camp, Deanne made a fire and I dressed for breakfast the way I had for dinner. Our request for fresh fruit was willingly and bountifully indulged, and we rolled up left-over bacon in leftover pancakes to supplement the avocados for lunch. The day soon warmed and we set forth up the trail to the next camp in shorts and T-shirts.

It never quite snowed on us and each day grew warmer, although I still needed my vapor barriers every morning and evening. Before the week was over we had made half a dozen new friends, supplemented trail travel with many miles of stimulating cross-country exploration, and climbed an 11,000-foot peak. And I'd caught and released forty to fifty Eastern Brook and Golden trout from half a dozen lakes and streams.

Every morning we enjoyed big bowls of fresh fruit, envied by our companions, some of whom vowed to try it. There was always enough left over for a second breakfast of oranges, raisins, and dates. By the end of the trip, the only clothing we hadn't worn were our Sierra Designs rain suits. We returned to Tuolomne, closing the circle, for a final night before departing, having enjoyed one of the world's premier extended (in our case nine-day) dayhikes.

Ours was a custom self-guided hike that cost about $60 a day each. The loop can also be negotiated as a seven-day guided tour by parties of up to twenty, with a park naturalist explaining the flowers, trees, birds, and geology. And there are guided four- and six-day saddle trips on mules for parties of ten.

I wish I could tell you that reserving a trip is easy, but the loop is also one of the world's most popular dayhikes. Many people apply year after year without success. One year I stubbornly made forty to fifty long-distance phone calls before finally securing reservations for three consecutive days at Vogelsang. But it was worth it. The ambience at the high camps—the spirit, the people, and the scenery—can't be beaten.

There are dozens of other areas throughout the world, some of them listed at the end of this book, whose operators and outfitters will take you on dayhikes of three days to three weeks, where you can savor a special beauty and where the experience doesn't end when the sun goes down. For Deanne and me, extended dayhikes are hiking at its finest, jewels in the everlasting memory of time spend walking in the wilds.

EIGHT

Homeopathy Revolutionizes First Aid

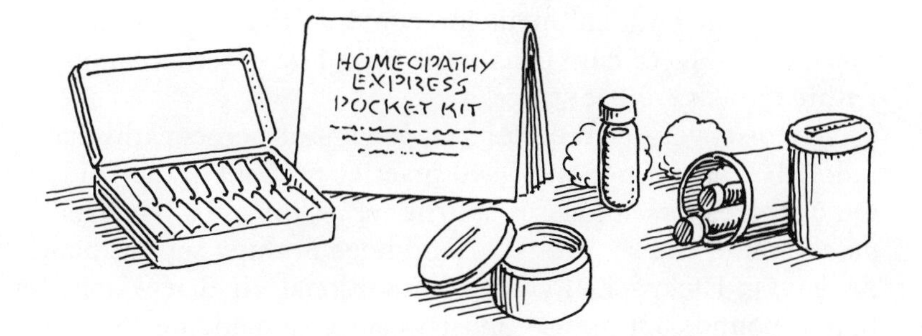

Homeopathic First Aid Kits . . . Safe Homeopathic Remedies Really "Cure" . . . How Homeopathy Works . . . Homeopathy Cures Almost Anything . . . What's in Those Little Kits? . . . Homeopathic Cures in the Wilds . . . Mending Broken Bones . . . Dr. Deanne Cures the Whitewater Crew . . . A Cure for Altitude Sickness . . . Another Cure on the River . . . The Doctor Treats Herself . . . Treating Shock & Bleeding . . . Minor Wounds . . . Sprains . . . Altitude Sickness . . . Sunburn vs. Tan . . . Burns . . . Hypothermia . . . Snakebite . . . Insect Bites . . . Foot Care Tips . . . Poison Oak & Ivy . . . Beware Sunglasses . . . Solo Travel Precautions

What if you could cure—really cure—nearly every type of ailment that threatens in the wilds with a featherweight first aid kit of safe, potent, inexpensive medicines? Well, you can. We've been doing it for years. These little known but increasingly popular medicines are called "homeopathic remedies." Here's an example of how well they work.

I was descending a Sierra peak when a wasp stung me on the finger. That may sound trivial—unless you've been deathly allergic to bee stings all your life. I was anxious about what might happen out there in the wilds, miles from a doctor. While I danced with pain, Deanne quickly fished a tiny vial from the kit in her daypack and put a single drop of Hypericum on the sting. It cut the pain in half in ten seconds. Five minutes later I couldn't remember which finger had been stung!

What conventional medicine could do that? And Deanne insists that Hypericum is only the third best external homeopathic remedy for bee stings!

My discovery of the curative powers of homeopathy has radically transformed the way I practice first aid in the wilds. On dayhikes I used to carry a little moleskin and a few band-aids for cuts and scratches . . . and hope nothing serious happened. On backpacks I carried a traditional kit that weighed half a pound, but it was mostly gauze and iodine. Now I carry one of several tiny kits of dynamite homeopathic remedies to magically cure a truly vast array of ailments commonly encountered on the trail.

Homeopathic First Aid Kits

About the size of a pack of cigarettes, our two-ounce kit, contains thirty different remedies and provides a comforting sense of security on major climbs and cross-country dayhikes (not to mention backpacks and river trips). On shorter dayhikes we take a featherweight five-remedy kit contained in two 35mm film canisters. On some trips these kits are supplemented by mechanical first aid equipment, like tweezers, bandages, razor blades, needles, etc. But all our medicines are homeopathic.

After five years of extensive research—during which I wrote a book on the subject and Deanne became a homeopathic practitioner—I've become convinced that homeopathy is infinitely superior to conventional (dark ages) Mountain Medicine. It's earned its reputation as "the most effective system of medicine known to mankind," even if it isn't widely known—yet.

Consider the range of ailments our little kits can treat: pain, shock, sprains, cuts and abrasions, altitude sickness, muscle strains, nettles, poison oak, insect bites and stings, sunburn, headaches, heat and sunstroke, bruises, choking, fainting, nausea, overexertion, blisters, burns, bleeding, diarrhea, shin splints, tendinitis and sore muscles, as well as other wounds and injuries, not to mention colds and flu. And our kits are highly flexible. We simply insert the remedies we're most likely to need for the journey we've planned in the kit we've decided to carry.

Safe Homeopathic Remedies Really "Cure"

When I say homeopathic treatment is "curative," I'm not exaggerating. This isn't the "temporary symptomatic relief" you're used to. And there's none of the wooziness or other dangerous side effects that we've come to associate with aspirin, antihistamines, Advil, Tylenol, codeine, etc. FDA-approved homeopathic remedies are so safe you can confidently treat babies—not to mention yourself, your friends, relatives, and companions. I've watched Deanne do just that with phenomenal success—on river trips, climbs, and foreign expeditions as well as hikes of all kinds.

Homeopathic remedies are wonderfully inexpensive. Our thirty remedy kit, including case and instruction booklet, cost only $32. We ordered it from a tiny ad in the back of *Backpacker.* It comes from Homeopathy Express, P.O. Box 213, Montchanin, DE 19710. You'll find the kit even more valuable at home between trips for treating colds and flu of all kinds, rashes, anxiety, earache, sore throat, dental pain, childhood illnesses, gas, constipation, boils, styes, impetigo, herpes simplex, ringworm, shingles, yeast infection, female problems, motion sickness, backache, and so forth. It will even work on your cats and dogs.

If homeopathy is so wonderful, you may justly ask, how come I've never heard of it? Fair question. Before presenting a selection of case histories from the wilds, let me quickly explain how homeopathy works, why it's so effective, and why you probably never heard of it.

How Homeopathy Works

Homeopathy's guiding principles go all the way back to Hippocrates in 400 B.C. But they were first systematized and medically applied two hundred years ago by a German physician named Samuel Hahnemann. He put the fundamental principle, "like cures like," to work. "Like cures like" means a medicine that will *cause* a certain set of symptoms in a healthy person—if taken in overdose—will *cure* that same set of symptoms in someone sick with them.

When homeopathy proved effective on the terrible epidemics that raged throughout Europe in the early 1800s, it jumped to America where it performed just as well, curing yellow fever, cholera, typhoid, and scarlet fever. By 1900, one American doctor in every four was a homeopath. There were twenty-two homeopathic medical schools, over one hundred hospitals, and more than one thousand homeopathic pharmacies. Homeopathy was successful—too successful for its own good!

The jealous medical establishment (doctors, their societies, and the drug industry) became alarmed at the fast-growing success of its upstart competitor. Homeopathy was too effective and too cheap. So war was declared. Using massive political power and money, the medical establishment managed to discredit homeopathy and cut off needed government funding. The system withered but didn't quite die—because it worked. And it rapidly spread throughout the rest of the world, where it wasn't unfairly persecuted by greedy big business.

But the good news couldn't be concealed forever in America. In the 1960s the public began to discover that miracle/wonder drugs weren't so wonderful after all. They were hideously expensive, they didn't really cure, and their disastrous side effects were filling up the hospitals. So Americans began to look for alternative forms of treatment, like acupuncture and homeopathy. By the 1970s, the homeopathic renaissance was well under way.

Now in the 1990s, homeopathy is booming. The word is out and the thinking public is beginning to demand it. Conventional medical schools still brainwash their students to believe that it's quackery and ineffective yet somehow dangerous. But the American public is fast discovering what the rest of the world already knows. Homeopathy works! It's safe, cheap, and best of all genuinely curative.

Conventional medicine depends on drug company concoctions dreamed up on a trial-and-error, hit-or-miss, basis—and hastily tested on rats. Its sole purpose is to devastatingly obliterate isolated unwanted symptoms, never mind the cost or disastrous side effects. By contrast, homeopathic remedies—painstakingly tested on healthy human beings—are prescribed according to a dependable principle (like cures like) to gently rejuvenate the patient's own defense and immune systems, thus effecting a safe and inexpensive cure.

Homeopathy Cures Almost Anything

Our application for first aid is only the tip of the iceberg. Homeopathy, when practiced by MDs in a professional clinic, is capable of curing chronic allergies, arthritis, heart disease, hypertension, gout, schizophrenia, depression and most other psychological afflictions, hereditary disorders, infectious diseases, skin problems, major accident damage, deeply embedded afflictions of every kind—even early stage cancer, multiple sclerosis, and AIDS! I've seen it happen.

I discovered homeopathy in the mid 1980s while searching for a cure for pollen allergies so severe that they kept me from living in my home in the desert. As my homeopathic physician (a Stanford MD) predicted, my allergy was cured completely in a period of three years. But some of the other benefits I received from this holistic (whole body) treatment were even more precious. To learn what it's like to be treated, and what homeopathy can do for you and your family, read my introductory book *Homeopathy: Medicine that Works!* ($9.95, paper, Condor Books). See Sources & Resources.

What's in Those Little Kits?

To illustrate homeopathy's effectiveness in the wilds I'm going to give you case examples. But to make them understandable, I first need to explain the contents of Deanne's Dayhiker First Aid Kit. Most homeopathic remedies come in the form of tiny treated sugar pellets packed in glass or plastic vials and taken orally. The rest are tinctures (liquids), lotions, or gels administered directly to the skin. The remedies in Deanne's sizeable home arsenal all come from Hahnemann Pharmacy in Berkeley.

Take a look at the kit we take on most dayhikes. Packed in a 35mm plastic film canister containing cotton balls (for padding and applications) are vials of Arnica (for shock, pain, bruises, swelling), Ruta (for sprains or pulled muscles) and Rhus toxidendron (for poison oak, tendonitis, strains). In a tiny fliptop vial is half a thimbleful of Hypericum tincture (for lacerations, stings, abrasions, severe burns, insect bites). Completing the kit is a small squeeze bottle of Calendula lotion, a first aid kit all by itself. Calendula is instantly soothing to the skin and heals a wide range of minor burns, abrasions, sunburn, superficial wounds, even athlete's foot.

Homeopathic Cures in the Wilds

Now for some case histories. Deanne and I were dayhiking the empty southern coast of Big Island Hawaii when I slipped on slick seaweed and fell full force on my face in the lava. Deanne opened our little kit and gave me Arnica, which reduced my pain by 50 percent within three minutes, eased my shock, slowed the bleeding, and lessened swelling. After washing in the sea I staggered back to the car where Deanne daubed my barely bleeding face with healing, disinfecting

Hypericum, the same remedy she'd used on my bee sting. During the hour and a half drive to our homeopath in Kona, my suffering was greatly reduced by repeated doses of Arnica.

Our doc said he couldn't improve on Deanne's first aid and sent us to a surgeon, who put a dozen stitches in my forehead. Then he wiggled my misshapen nose and pronounced it unbroken. When his nurse offered pain pills, I said no thanks, I had my own. When she tried to swab my wounds with iodine, Deanne stopped her. Iodine, she knew, burns the epithelial tissue of the skin, leaving it vulnerable to infection, hindering healing. When we got home, Deanne kept my wounds clean with more Hypericum.

Mending Broken Bones

When scabs quickly formed, she switched to Calendula to promote rapid healing while keeping the skin soft. When we returned to the surgeon to have the stitches removed, he was impressed by the lack of swelling and the swiftness of healing. Again he tweaked my tender nose and assured me it wasn't broken. My Frankenstein forehead was completely healed and scarless within two months, thanks to Arnica, Hypericum, and Calendula treatments, but my nose was still so tender that I cringed whenever Deanne kissed me.

"Your nose *has* to be broken," she frowned. "I'm going to give you Symphytum. They call it 'Knitbone.'"

After just five daily doses, the tenderness was entirely gone and I could wiggle my nose without pain. Knitbone had done the job. I was impressed. Two months later, our friend Yon told us the doctor was about to put his leg in a cast for months to heal a cracked knee bone. Deanne begged him to try Symphytum first. He did. After five daily doses the X-rays showed Yon's knee to be completely healed. He cancelled his appointment with the doctor.

On a six-day raft trip down the Salmon and Snake rivers in Idaho the following summer, Deanne brought her eighty-remedy home kit because our outfitter friends had announced that she'd be willing to treat any of the seventeen members of the crew. We were hardly under way when a boatman named Ricardo, also allergic to bee stings, was stung by a hornet on the back of the neck. Deanne put a single drop of Ledum on the red-bordered white welt the size of a dollar. A look of disbelief came over Ricardo's face. Half his pain was gone in seconds. That night when we pulled ashore to make camp, the welt was gone. The Ledum had accomplished in four hours what usually took Ricardo four long painful days.

Dr. Deanne Cures the Whitewater Crew

Our river companions were quietly impressed. Before long they began to bring Deanne their problems. First came Tom with a badly abraded hip. Calendula brought immediate relief, just as it did for Karen's sunburn. The next day, Deanne gave young Teddy Rhus tox for the swelling and eruption of poison oak on his arms. By sundown the itching and swollen rash were gone.

Teddy's dad, Lew, complained of intense pain and swelling in his back after sitting in a cramped raft and paddling all day. Just a month before, he'd had spinal surgery to fuse two disks. Now he feared that he'd made a mistake in coming on a wilderness raft trip so soon. Deanne gave him four doses of Arnica for the pain. When it was nearly gone, she switched to Rhus tox for strain. Relief from the two remedies kept Lew happily paddling the rest of the trip.

Trip Captain Judy brought Deanne a painfully swollen finger. It throbbed with pain as she paddled for five hours every day. Together they opened the finger, and Deanne treated it four times every day with Hypericum. Like Lew, Judy was able to keep on paddling as her finger gradually healed.

Before the trip was over, more than half the crew of seventeen came to Deanne for homeopathic treatment of some kind—and that included us. When Deanne bit open the inside of her cheek, she healed it quickly with Calendula lotion. When I was catapulted from my seat at the oars in a rapid, she treated my barked shin with Arnicated oil. An hour later, the pain and bruise had disappeared.

A Cure for Altitude Sickness

On a high Sierra pack trip with the Sierra Club near Yosemite, one of our leaders, the wife of the trip doctor, complained that altitude sickness at 11,000 feet was keeping her nauseous, preventing her from eating and sleeping. The Advil her husband had given her hadn't helped. She was miserable and getting weaker, and ready to try anything. Deanne's homeopathic remedy cured her completely overnight. And the doctor became interested in homeopathy.

That winter, in a steady downpour, we began a six-day hut-to-hut tramping trip in New Zealand's wild rain forest. Before long we were soaked to the skin, and by evening we were chilled. Not surprisingly, we awoke the next morning with savage colds that threatened our enjoyment of the country we had come so far to see. Outside the hut it was still raining. Before putting on damp boots and setting forth, Deanne studied our symptoms, consulted her manual, and dosed us with what she hoped were the right homeopathic cold remedies.

By noon, when we reached the pass, the sun had come out and our cold symptoms were dwindling. By evening, despite more rain, our colds were entirely gone and we felt fresh and strong, ready to savor the rest of the trip. We weren't surprised. The right homeopathic remedy regularly obliterates colds, with no lingering symptoms and weakness. That's the different between "cure" and mere "temporary symptomatic relief."

Another Cure on the River

Last June I was captaining a raft down the 225-mile Grand Canyon when a violent rapid folded the heavily laden boat in two, smashing my right foot with an ice chest. The sharp pain suggested that something was broken—and we had another hundred miles of river to navigate before reaching civilization. Deanne had a dose of Arnica in my mouth within seconds, cutting my pain in half. Then she made me hang my foot over the side in the 45-degree water while she packed ice from the chest in a plastic bag to make a cold compress.

That night I crawled off the boat at our camp on a sandbar. Deanne washed my wounds with Hypericum and switched from Arnica to Bryonia, coating the unbroken skin of my foot with Arnica Gel. The relief from pain allowed me to sleep.

The following day, thanks to her treatment, I was able to guide my boat through another day of rapids. The sharp pain was gone and I was able to finish the trip at the helm with only mild discomfort. A week later, a combination chiropractor/homeopath diagnosed a dislocation in my foot and reset it. Several weeks later, after too much hiking, it dislocated again. This time when he reset it, he gave me a dose of Rhus tox. After two days of rest the foot was fully healed and I was ready for our week-long, camp-to-camp dayhiking trip in the high Yosemite backcountry.

The Doctor Treats Herself

But two days before our departure, Deanne twisted her knee. It hurt with every step. A quick dose of Arnica promptly stopped the pain, then she switched to Bryonia while locally applying Arnica Gel, the same prescription that had helped me on the river. Her knee was fully recovered by the time we left for Yosemite.

This handful of sample cured cases from the wilds should explain why nowadays we rely for all our medicinal needs on homeopathic remedies. No traditional medicines can match them for safety and effectiveness. And they cost only pennies per dose. Of course wilderness first aid—truly first care—generally precedes the taking of any medicine. But there isn't any conflict between true first aid and the use of homeopathic medicine. Far from it. Homeopathy admirably complements first aid techniques of every sort and makes them infinitely more effective.

Take snakebite, for example. Instead of relying on the old discredited prescription of "cut, suck, tie"—and suffer, homeopathic remedies like Lachesis and Ledum work wonderfully to internally counteract the effects of the venom. It's

the same with wounds and injuries of all kinds, shock, bleeding, pain, swelling, and so forth. We still carry a traditional first aid kit when the situation requires it. But when it comes to medicine, we rely on homeopathy—because it works!

If you want the greatest possible protection, the capacity for curing ailments of nearly every kind, get yourself a kit and a book and learn how to use them. Your increased capability will give you the same peace of mind we enjoy in the wilds, at home, and wherever we travel. It can save you a lot of suffering. It might even save your life!

Now let's turn to the recommended first aid treatment—both conventional and homeopathic—for the various afflictions that commonly beset the dayhiker.

Treating Shock & Bleeding

Shock is a state produced by injury or fright. The victim feels cold and clammy and weak. The treatment is to lay the patient down on level ground and make him as comfortable as possible, usually by loosening constricting clothing and covering him if it is cold, until a feeling of well-being returns.

In case of a small or slightly bleeding wound, bleeding usually will soon stop if the wound is elevated so it lies higher than the heart and pressure is applied with a gauze pad. (For a cut foot or leg, the patient lies down and props his leg against a tree; a cut hand should be held above the head.) A large or heavily bleeding wound may have to be closed by hand pressure. A puncture can be firmly blocked by the palm or a finger. On a slice or cut it may be necessary to draw the edges together with the fingers before applying pressure. Closing the wound to stop the bleeding is vital. Once bleeding has been controlled, the wound should be kept elevated to reduce the blood flow and aid clotting. Never attempt to substitute a tourniquet for these procedures.

As soon as bleeding is under control, the wound should be washed with soap and water, or irrigated with water, to carry away bacteria and dirt. It may be necessary during the washing to keep the wound elevated to lessen bleeding. Once cleaned, it may be gently blotted dry with a clean cloth or towel (not to mention toilet paper or clean socks). The clean, dry wound can then be bandaged. On heavily bleeding wounds that do not respond sufficiently to elevation, it may be necessary to tape the edges of the wound together with a butterfly bandage in order to stop bleeding.

Homeopathic Arnica is specific for shock, bleeding, and swelling. Given orally in the appropriate potency, Arnica immediately reduces pain and shock, swelling and bleeding by an average 50 percent. We don't set forth without it.

Minor Wounds

Minor cuts and scratches, especially on protected parts of the body, are better left unbandaged. Protected but uncovered wounds are more easily kept clean and dry; healing is faster and the chances of infection are lower. Antiseptics (Mercurochrome, iodine, methiolate, and the like) should not be applied; they tend to do more harm than good—inhibiting scab formation and trapping bacteria which cause infection. Small wounds need only bandaids. Larger ones will require a gauze pad held in place by narrow strips of adhesive tape. The largest may require wrapping the limb or body with roll gauze. Gauze and adhesive bandages should be applied directly on top of a wound held closed by a butterfly bandage.

The greatest enemy of wounds is dampness. A wet bandage inhibits healing by providing a favorable environment for the growth of bacteria. Once a bandage has become wet, whether from blood, perspiration, or water, it is a menace to health and should be replaced. No bandage at all is far superior to a wet one. The drier the wound the less the chance of infection.

Homeopathic Calendula, lotion or tincture, is wonderfully soothing and promotes healing on superficial wounds of all types. Hypericum tincture is superior for deeper wounds and Ledum is best for punctures. Calendula is a first aid kit all by itself.

Soap and plenty of water are better than antiseptics and first aid creams, and tourniquets are dangerous and rarely essential since pressure and elevation should stop all but the most serious bleeding. Amateurs should never attempt to set broken bones, but splinting and immobilizing breaks is part of first aid. Seriously injured patients should be taken out to a doctor if they can travel; otherwise bring the doctor in.

Immediately taken orally, high-potency Arnica is a godsend for the pain and shock of broken bones, and it holds down swelling in the surrounding tissue.

Sprains

Nothing is more common among dayhikers accustomed to doing their walking on sidewalks than turned or sprained ankles. Severity varies greatly. Some sprains amount to nothing more than a momentary twinge. Others require the victim to be immobilized immediately. Often the wisest course for the person who has suffered a bad sprain (the ankle immediately turning black and blue) is to apply a tape cast and head for the car before the ankle can swell and stiffen.

sprained ankles can be chilled in an icy stream or with snow-filled plastic bags

Moderate sprains should immediately be treated with cold to constrict blood flow and prevent swelling. Putting the foot in an icy rill or applying cold compresses made by filling plastic bags with snow or ice water are fast and effective. Elevating the ankle also helps greatly to reduce the swelling. If sources of cold are not handy or it is inconvenient to stop, an elasticized ankle brace of three-inch Ace bandage may be applied.

Braces are likely to be carried only by people with weak ankles who have come to rely on them. Ace bandages have the advantage of being usable on other parts of the body. In either case, it may be necessary to remove all (or at least the outer) socks to make room for the bandage in the boots. And people (like myself) who have sensitive Achilles tendons may find it impossible to wear an elastic bandage very long. Bandages need only be worn while walking. They should be removed at night and at any other time that the ankle can be elevated.

All of the swelling that is going to take place will happen on the day of the sprain or the day that follows. On the third day, with the swelling stopped, the treatment changes from the application of cold to the application of heat. The intent now is to stimulate blood flow through the injured area in order to reduce swelling. Hot compresses made from bandanas, towels, diapers, or washrags dipped in heated water are excellent, or the ankle can be baked before an open fire. Hot water bottles can sometimes be fashioned from large plastic bags, but care must be taken not to burn the patient. The exception to heat treatment is the ankle which is immediately encased in a cast of tape. Such casts should be left undisturbed for two or three days and heat applied only after removal.

The homeopathic treatment for sprains and strains starts with Arnica for pain and shock, followed by Rhus toxidendron if the injury feels better with movement, or Rut if it's worse with motion.

Altitude Sickness

As altitude increases the oxygen content of the air decreases. In order to adjust, the body strives to process more air by means of faster and deeper breaths, to better extract oxygen from the air. Adjustment begins at only slight elevation, but shortness of breath and dizziness do not usually appear until about 7,000 feet. Individual tolerance to altitude varies widely. The more gradual the change in altitude, the easier the acclimatization. The well-rested, vigorous, healthy individual usually acclimatizes easily. Smoking, drinking, and heavy eating before or during a climb make acclimatization difficult.

Failure of the body to adjust to reduced oxygen intake results in "altitude" or mountain sickness. Mild symptoms include headache, lassitude, shortness of breath, and a vague feeling of illness—all of which usually disappear after a day of rest. Acute mountain sickness is marked by severe headache, nausea, vomiting, insomnia, irritability, and muddled thinking. The victim must descend to a lower elevation. Mountain sickness can usually be avoided by beginning a trip in good condition, spending a night at the trailhead before starting out, and choosing modest goals for the first day's walk. Most acclimatization occurs in the first two or three days. The rule of thumb calls for "climbing high but sleeping low."

Highly dilute, homeopathically prepared Coca cures altitude sickness. Though harmless it can't legally be sold in the United States.

People who acclimatize poorly, when they reach elevations in excess of 10,000 feet, are susceptible to high altitude pulmonary edema (HAPE) (fluid accumulation in the lungs). The first symptoms include a dry, persistent, irritating cough, anxiety, and an ache beneath the breast bone and shortness of breath. If the victim is not evacuated promptly to lower elevation or given oxygen, breathing may become rapid, noisy, and difficult, the skin often takes on a bluish tinge, and death may occur quickly.

Sunburn vs. Tan

Sunburn is a constant threat, especially at higher altitudes, to city dwellers who are not deeply tanned. At 6,000 feet the skin burns twice as fast as at sea level, and the liability continues to increase with altitude. Sunburn often ruins a trip when a pale backpacker tries for a fast tan. Precautions should be taken to cover—or at least shade—all parts of the body for most of the day. Few people ever acquire a deep enough tan to expose themselves all day at high altitude without burning. Special care should be taken to avoid burning the nose to prevent starting a cycle of peeling, burning, and repeeling.

Nothing is worse than having to hike in the heat completely shrouded from the sun—unless it is suffering with sunburned shoulders that will have to carry a pack the next day. Trying to safely tan white skin by short periods of exposure to fierce high altitude sun is a bothersome, inconvenient process. Potent (#15) sun screen ointment is essential on most trips. So is protection for the lips. The best treatment for sunburned skin and other first-degree burns is Calendula, which is marvelously soothing and prevents peeling.

Burns

For first- and second-degree burns, apply cold water for five minutes or until the pain stops. Then keep burns dry, avoiding oils and greases. Homeopathically, use Hypericum externally for second degree burns and Cantharis orally for third degree burns.

Hypothermia

The number one killer of outdoor travelers is hypothermia, defined as "rapid mental and physical collapse due to chilling of the body's core." When the body loses heat faster than it's being produced, you instinctively exercise to keep warm while the body cuts back blood supply to the extremities. Both drain your energy reserves. If chilling and exposure continue, cold will reach the brain, depriving you of judgement and reasoning power without your awareness. As chilling progresses, you lose control of your hands and body. When your body can no longer summon reserves to prevent the drop in core temperature, stupor, collapse, and death await.

The first line of defense is awareness, awareness that most hypothermia cases occur during mild temperatures, 30 to 50 degrees Fahrenheit. The greater hazards are wind and wet. Wind drives away the skin's cushion of warm air, and it refrigerates wet clothing. Remember that 50 degrees Fahrenheit water is unbearably cold, and that the wet body can lose heat two hundred times as fast as one protected by dry clothing! There is no better clothing for hypothermia protection than vapor barriers. If you can't stay dry and warm, do whatever is necessary to stop exposure. Turn back, give up, get out—before exhaustion can complicate your plight. Don't shrug off shivering. If you have to exercise continuously to prevent it, you're in danger. Get out of wet clothes, get dry, and put on vapor barriers to stop heat loss, then take hot drinks, heap on the insulation, utilize whatever heat sources are available, and stay awake.

By paying attention to what your body tells you, by proper use of your clothes, and by taking precautions against injury, you can double your chances of enjoying a safe, happy trip.

Snakebite

First prerequisites in prevention are caution and the ability to recognize poisonous snakes and the sort of terrain they like. I have spent a good deal of time in heavily infested areas and have encountered a great many rattlers. But by never extending any part of my body into a concealed place that could contain a snake, I have avoided being bitten. While fatalities from snakebite are rare, travelers in poison snake country need to be prepared.

Since the old "cut, suck, tie" formula was discarded as dangerous for the sucker and ineffective for the suckee, there has been no agreement on first aid treatment. Carrying antivenom is impractical. Some say get the patient to a hospital, pronto. Others stress immobilization and ice, but all are opposed to using tourniquets. If symptoms are severe, see a doctor.

As indicated earlier, homeopathic Lachesis and Ledum are effective. So is Carbolic acidum, depending on the patient's particular symptoms.

Insect Bites

Biting insects are everywhere and threaten the enjoyment of many a trip. Mosquitos, spiders, gnats, black flies, noseeums, sand flies, etc., are hard to defend against. Heavy clothes, headnets and a coating of repellent aren't always sufficient or convenient.

If you need bug repellent in quantity it makes sense to buy diethyl metatoluamide at the drugstore and make your own dilution. A dosage of 200 milligrams/day of vitamin B-1 taken orally will make your perspiration repellent to mosquitos and thus keep them away. So will the heavy consumption of garlic.

Bug bites can be homeopathically treated with great effectiveness by taking Apis or Ledum orally and by applying Hypericum or Ledum locally for itching, pain, and swelling.

Foot Care Tips

A backpacking doctor says, "At the first hint of discomfort, stop, take off the boot, and have a look. Wash and dry a place that is getting red, then tape a thin sheet of foam rubber over the spot." I had always relied on moleskin for covering blisters and inflamed places on my feet. Moleskin's disadvantage is that once it is stuck directly to the injured or tender area it cannot safely be removed (without removing the skin) until the end of the trip. In the meantime, of course, the moleskin is certain to get damp and dirty, encouraging bacteria growth. On his advice I have switched to either Molefoam or foam rubber and find both perfectly satisfactory.

Often as important as bandaging an inflamed foot is attacking the cause of the inflammation. On occasion I have had to hammer down a nail with a piece of granite or whittle away a protruding ridge of leather. More often the problem is solved by kneading new boots that pinch, removing a pebble, loosening laces, removing the wrinkle from a sock, adding an extra pair of socks, or changing to a dry pair.

Long toenails will make boots seem too short and can be painfully crippling on downhill stretches. But cutting long toenails the night before a trip will result in pain and inflammation on the trail. Soothing Calendula, applied after washing, will heal blisters, hot spots, chapping, and raw skin—anywhere on the body.

Poison Oak & Ivy

Poison oak, like rattlesnakes, is a hazard that can usually be avoided by caution and the ability to recognize the danger. Poison oak in the west and poison ivy in the east have oily-looking, distinctive three-lobed leaves that are easily remembered once they have been identified. Tolerance to the oil, which remains potent for some time on clothes and on the fur of pets, varies widely. Persons exposed have a second chance to avoid the itching, easily spread rash by scrubbing exposed skin vigorously with soap and hot water on the same day. Skin irritation generally begins four to five hours after exposure. In the west, poison oak rarely grows above 6,000 feet.

Homeopathically prepared Rhus toxidendron, taken orally, is specific for poison oak and ivy.

Beware Sunglasses

It isn't widely known that sunglasses often do more harm than good. They trick the eyes into staying open wider than they should in bright conditions, resulting in eyestrain. And the darkened lenses block out healthful rays that are essential to the body. A wide-brimmed hat or sunshade is always preferable. "Dark glasses are a crutch," said the old prospector with whom I used to travel on the desert. "Put them on when it's bright and you'll never take them off." He taught me to squint and wear my hat low for a couple of days to acclimate my eyes, rather than develop a dependency on shades. And it works. I only wear sunglasses now under extreme conditions.

Murl also taught me to use "Indian sunglasses" when vision was vital under extra-bright conditions. Put the tips of your middle fingers together, end to end, then tuck the tips of your index fingers together tight against them, just beneath. Hold your four fingertips against your nose in the hollow beneath your brow and look through the easily adjustable slits between your fingers. Now your shaded eyes can stop squinting and open wide for maximum vision, even when looking almost into the sun.

Salt pills (five grain) are not required by most people unless the perspiration is literally pouring off the body. The usual dosage in such cases is one pill every four to eight hours, but only while drinking a quart of water per one to two pills. Overdosing on salt is dangerous!

Solo Travel Precautions

From a safety standpoint, the greatest danger in wild country is traveling alone. Unfortunately, some of the joys of wilderness travel are only to be discovered by traveling alone. When I hike by myself, I tell someone responsible where I am going, what route I plan to follow both directions, when I expect to be back, and the latest time (the time to begin worrying) that I could possibly be back.

In conclusion, let me urge you again to try homeopathy. It routinely doubles the effectiveness of first aid and can save much suffering—perhaps even your life! It's safe, inexpensive, and easy to use. The more you travel the wilds, the more valuable it will be. A small kit and a manual could be the best investment you ever made.

NINE
Trailside Tips & Techniques

Warm-ups & Cool-downs . . . What to Wear . . . Pacing Yourself . . . Walking Dynamics . . . Uphill & Downhill . . . Group Walking . . . The Art of Resting . . . Family Walks . . . Motivating Little Hikers . . . Toilet Training . . . Cut Your Toenails . . . Get Off the Trail . . . The Joys of Climbing . . . Baby Your Tootsies . . . Walker's Enemy: Chafing . . . Test Your Traction . . . The "Limp" & "Indian" Steps . . . Fording Streams . . . Use Those Arms . . . Daydream Painkiller . . . Trail Manners . . . Cleaning Up the Country . . . Safety Tips . . . Reading the Weather . . . Storm Warnings

Over the years, a veteran dayhiker like myself discovers a strategy or two for making his walks easier, safer, more comfortable—and above all more fun. In hopes that my experience will do the same for you, I've compiled a distillation of the specific ways that I cope with the vicissitudes of walking.

For a quickie or fast aerobic walk, what you wear—and *don't* wear—can be surprisingly important. I've learned the hard way to *under*dress, especially if I plan a brisk pace. I try to deliberately start out feeling chilly. If I don't, before I'm done I'll be sweating and wondering how to carry all the damp clothes I've been forced to peel off.

Warm-ups & Cool-downs

It can be vital to loosen, limber up, stretch, and otherwise warm up your muscles before strenuous exercise like dayhiking—especially if you're fresh from watching TV. You need to work the rust from those hinges and oil them, slowly. It can be dangerous to subject the body to sudden stress. Sudden action can cause stiffness, soreness, early fatigue, torn tendons, and pulled muscles. The older you are or the colder it is, the more important it is to warm up.

It has been said that slow walking—half to three-quarter speed—is the best warm-up for walking. But I always stretch the backs of my legs (hamstrings) and Achilles tendons by bracing my hands against a tree or rock, straightening my body at a forty-five-degree angle to the ground, and gingerly lowering my heels to the ground. Over the years I've added other exercises from my experience in yoga, tai chi, and dancercise and from conversations with chiropractors, runners, and physical therapists.

I start with traditional tai chi warm-ups: slowly flexing the wrist, ankles, elbows, and knees, circling the head on the neck, circling each shoulder, windmilling the arms, finishing with slow-motion bent-knee lunges with my arms outstretched. By adding conventional exercises, like touching my toes, I endeavor to gently stretch all my muscles and open all my joints.

My last stretch before hitting the trail consists of standing on one leg and pulling the foot of the free leg all the way up to my buttocks. It's also important, especially if it's cold, to freshly stretch the Achilles tendons after each stop or cool-down. Warm-ups aren't just for the overcautious, the aged, or sissies. If you're not properly warmed up, a misstep can cause a muscle pull, a slip can mean a strain, a fall can result in a sprain. And repeated injuries can lead to painful chronic problems like arthritis and bursitis when you're older.

During a hard hike—especially after a long, jarring, downhill stretch—when I stop to rest I lie down on my back and elevate my feet by putting them higher than my head against a boulder or tree. And I hold my hands in the air. The object is to redistribute the blood that's been hammered down into my hands and feet, hopefully back to my heart and brain. It's a little like turning over an hourglass. Within seconds I can feel my feet begin to shrink inside my boots and the swelling in my hands subside.

It can be just as important to cool down gradually, to avoid the sudden changes that stress the body and can make you dizzy. The harder the workout, the more important the cool-down. A couple of minutes of slow walking will usually suffice, but if the hike has been grueling or my pack was heavy or I'm stiff and weary, I find it wonderfully stress relieving to systematically loosen my muscles after hiking. Five minutes of yoga stretching, culminating with a headstand, marvelously rejuvenates and refreshes me.

The transformation must be experienced to be believed. When ligaments and muscles are gently guided back into their natural state after prolonged effort, stiffness and fatigue simply dissolve. The single most rewarding exercise for me after hard hiking involves lying on my stomach, putting my hands in the push-up position, then, keeping my pelvis on the mat, gradually straightening my arms, throwing back my head, and bending my spine backward, keeping my knees locked.

After several deep breaths, I turn my head as far as possible (until I can see my feet) on one side, then the other. Then I relax my knees, lower my elbows to the pad, and lie relaxed, back still bent, breathing deeply for another minute. Other stretches of great benefit include slowly rotating the neck, rotating the shoulders, and rotating the ankles.

What to Wear

If possible, I wear shorts. My legs are hairy, and I hate to overheat, so I don't bundle up in even the lightest of long pants unless the temperature is in the forties. I learned long ago, hiking on ice and snow in the Swiss Alps, that I won't freeze in shorts so long as my trunk and head are well protected. So, for a quick walk, I wear shorts and a thin shirt, nothing else in warm weather. If it's cool or windy I might add a wind shell. If it's really cold I might substitute a sweatshirt for the thin shirt.

But I expect to return with the shell rolled up and tied around my waist by the sleeves. And if I really get warmed up, the shirt will be tucked into the back of my shorts, hanging and swinging like a tail. I carry nothing whatever in my hands and no more than a handkerchief and lip ice in my pocket. The only possible addition to my wardrobe for a quickie is a hat, or more likely a cap, to shade my eyes from the sun.

For longer walks I may have to take a pack, if only to accommodate all the clothes I take off. I don't take one unless I have to, and my first choice is a small belt- or fannypack. Since I'm taking a pack, it means the weather is extreme and I need to bundle up or I'm going to get hungry and need to carry food. If I'll be back, or somewhere else, for lunch, I'll take along fruit, probably a peeled orange and grapes if water will be short.

If water is plentiful and the walking is serious or steep, I'll probably rely on dried bananas and dates. Either way, I may include a granola bar as backup in case I'm late for lunch. And on casual walks I never carry water. If I'm afraid I may get thirsty, I imitate the camel. I drink a glass of water, no more, before I go—ideally half an hour before. In fact, if I'm going on an all-day hike, I start drinking water as soon as I get up. I want to start my walk well hydrated but not sloshing. If the hiking will be warm, I slightly underfill my water bottle the night before and put it in the freezer. In the morning I top it off, looking forward to sipping ice water beneath the broiling sun.

Pacing Yourself

There is a myth that one should find a comfortable pace and then stick to it. Nothing could be further from the truth. The most common error among hikers is trying doggedly to maintain a set pace despite changes in the grade. Constant speed is an impossible goal. Comfortable, efficient walking depends on maintaining one's energy output—not one's speed—at a level that will not produce excessive fatigue. This simply means slowing down when the trail climbs, then speeding up when it levels off.

The length of one's stride should also be variable. When the trail suddenly grows steeper, I not only slow down, I take shorter steps. When the trail levels off, my stride gradually lengthens. Walking in this manner, i.e., trying to maintain an even and comfortable output of energy rather than trying to maintain a constant speed, I am never forced to stop from exhaustion, and I log more miles per day in greater comfort.

There's a scientific basis for my "variable speed" philosophy. Specialists have determined that for every individual and set of walking conditions there is an *ideal* pace, an optimal speed that requires minimal energy per step. Our internal computer instinctively tries to conserve energy and it will govern our speed for maximum efficiency—if we let it! Ignoring the clockwork within ourselves by hurrying and even by moving too slowly will be far more tiring than the optimal pace because it uses more energy than necessary.

Walking Dynamics

The experts have learned some interesting facts that can help us walk more functionally. Walking involves about a hundred different muscles, but *all* the walker's propulsive thrust is delivered by the terminal bone of the big toe. Our computer propels us by converting potential energy to kinetic energy with almost 50 percent efficiency. Walking is a state of carefully controlled falling, using the acceleration of gravity for the purpose. There is an advantage, it turns out, to a certain amount of bobbing as we walk. The extra work of raising the body increases the help we get from gravity by permitting us to fall farther with each step. Walking downhill is easier because the body can fall farther. Freely swinging arms help walking efficiency by stabilizing the shoulders and pelvis with the thrust of their counterrotation.

Hurrying, especially uphill, can be counterproductive in another way. Superexertion produces lactic acid in the blood, which hampers muscle performance, causes great discomfort, and requires more than an hour for recovery, during which the walker suffers from exhaustion. So the clever hiker's strategy requires keeping activity level below the lactic acid formation stage. On difficult grades that means slowing the pace to a comfortable level or stopping frequently to rest and allow oxidation to flush the blood of acid buildup. By experimentation I have discovered that on the steepest trails I drop below the level of painful lactic acid buildup by shortening each step from eighteen inches to fourteen. That four inches makes a huge difference in my comfort.

Uphill & Downhill

It is important to react immediately to changes in grade. Failure to cut speed instantly when the trail turns abruptly upward places a demand on the body for extra exertion. And extra exertion consumes a disproportionately large part of one's store of energy. For instance, with the energy required to run fifty yards uphill one can easily hike a quarter mile up the same grade in far greater comfort. Large expenditures of energy—running, lunging, jumping, taking huge steps, even hiking too fast—must be avoided.

On a really steep slope, at high altitude under load, or where the footing is bad (sand, scree, or loose snow), I adjust my pace even more precisely by controlling my step-to-breath ratio. I may, for instance, take two steps to the breath, inhaling as I plant my right foot and exhaling as I plant my left. If that proves hard to maintain, I may slow to a breath for every step or even two breaths per step, with a greatly shortened stride. On exceptionally difficult slopes it is better to slow to a crawl, taking six-inch steps, than to make the frequent stops a faster pace would require. Starting and stopping consume extra energy. A dependable rule of thumb is that where the going is hard it is better to slow down and keep going than it is to make frequent stops. An unlooked-for dividend of step-to-breath counting is the welcome distraction the counting provides.

Having dealt with uphill and level trail hiking, it is time to go down. It is common to feel relief when the trail starts down because it is so much easier on the lungs. But downhill travel is twice as hard on the legs as going up. When descending a steep trail I try to cushion the shock of each downward step by rolling my hip forward (not unlike the Indian Step movement described later in this chapter) and placing my foot with the knee slightly bent. As I transfer my weight I allow my knee to flex so that it functions in much the same fashion as an automobile shock absorber, reducing the jarring that downhill travel inevitably produces.

When I'm walking with my wife, I find it difficult not to keep in step with her. There's something about the rhythm and cadence—and the compulsory marching when I was in school—that drives me to match my step to hers. When I unconsciously fall into step, I soon regret it. To keep abreast of her and keep in step I have to shorten my stride uncomfortably. Deanne is five inches shorter than I am, so her legs are at least two inches shorter. As a result, she has a natural stride that's a few inches shorter than mine.

So, if we're going to walk together, she has to speed up or I have to slow down. Since I strongly believe that everyone should be free to travel at his or her own speed, sometimes we part. I may walk ahead, then stop to wait for her. She likes to walk energetically, and she's twenty years younger than I am, so we travel uphill at close to the same speed. I'm a bit faster downhill, so we often separate if the way down is difficult, reuniting at the bottom.

Group Walking

The fact that people walk at different speeds is neither bad nor good. Don't feel guilty if you're slow and don't brag, judge, or blame if you happen to be fast. For happy carefree walking that doesn't tire you out, protect the integrity of your own individual stride. Walking with the brake on can be tiring because it's unnatural. Trying desperately to keep up with longer-legged companions is even worse. I'm intimately familiar with both positions.

The best compromise involves stopping periodically if you're fast to let your partner(s) catch up. And don't dash off the moment they get close. Being acutely aware that they're holding you up, they're probably pushing themselves a little, so they could use a companionable moment or two to blow and visit before you take off and they fall behind again.

On group hikes the slowpokes often get left far behind. If they're also inexperienced walkers and don't know the country—very often the case—an assistant leader content to poke along should be detailed as "rear guard" or "sweep" to be sure no one gets lost or left behind. Larger groups will often sort themselves out into a fast section and a slow one, allowing the members of each to stay comfortably and companionably in sight of one another.

Of course social considerations may dictate considerably more togetherness. Walking with a partner or a group can be the very best form of companionship. The miles slip away when you're talking to a friend. And there's something about walking together that invites intimacy. Maybe it's the stimulation of motion and the relative freedom from interruption. There isn't any awkwardness because you're doing something together. The country rolls by like a slow moving picture. There's no doorbell, phone, or TV to interrupt. If you want to really get acquainted with someone, take them for a two-hour walk in the country.

Where togetherness isn't vital I urge people to travel at their own pace, the tortoises starting earlier and meeting the hares at the halfway point for lunch. I also advise people to savor the joys of solo travel or pick companions with similar capabilities. Increasingly on the trail I meet carefree groups of women happily poking along together, free of the strain of performing or keeping up with men.

The Art of Resting

Despite my advice to slow down and keep going, rest stops are a vital part of walking. They offer a means of savoring the country as well as restoring the body. One school holds that rests ought to be ruled by the clock, i.e., so many minutes of resting followed by so many minutes of hiking. This arbitrary arrangement makes no allowance for the difficulty of the terrain or the allure of the country. But what's worse is the notion that one needs to be ruled by the clock, even in

Rest stops are a vital part of walking

the wilderness. The tyranny of time, it seems to me, is one of the things that people go to the woods to escape. I am willing to admit the usefulness of a wristwatch in the woods for arranging a rendezvous with other watch-wearing members of the party, but I find clock time as dispensable in the wilds as doorbells, radios, telephones, and cars, and I refuse to carry a watch, with only minor inconvenience.

Getting back to rest stops, most walkers, provided they have a modicum of self-discipline and know how far they have to go, will find it more satisfactory to rest when they want to or need to. I like to stop, if I can manage it, beside a stream, at the top of a slope, in the first shade after a treeless stretch, where a log or rock forms a natural seat, or at any point where the view is unusually fine. I also favor mossy dells, waterfalls, brilliant patches of wildflowers, and fords where I can wash my feet or set up my rod and take a few casts.

When it comes to a real rest, I like to imagine I have earned it. On a particularly difficult slope, for instance, I might promise myself a rest after another hundred steps. Sometimes a hundred is impossible and I have to settle for fifty or even twenty-five. But if I get to thirty-five and think I can squeeze out another fifteen I try it. For variety and to add to the distraction, I sometimes count my steps backwards.

When I am ready to rest I take some pains to enjoy it. I slip out of my pack and sit or lie down. If my boots are the least bit uncomfortable or my feet are damp, I take off both boots and socks and set them to air in the sun or breeze. If there is water running nearby, I give my feet a soapless washing and rub and let them dry in the sun. If I am feeling faint or tired I lie down with my feet propped high against a tree so the blood can drain from my legs back into my body. Once my fatigue has drained away and my breathing has returned to normal, I usually have something to eat.

Sometimes the greatest benefit of a rest stop is having some fun, doing a little exploring. I like to stroll away from the trail to have a look at country I would otherwise miss. Often enough, I discover something unsuspected: an abandoned prospect hole, a bed of mushrooms, a hidden view, the remains of a lean-to, a tiny spring, or a wild sheep horn.

A rest may last anywhere from thirty seconds to an hour. When the time comes to move on, it is vital to start out at a moderate pace. There is a tendency to rocket up the trail after a refreshing rest. I have often seen eager kids start off at a run, slow to a walk, then sink into a panting, dispirited trudge—all within sixty seconds. So, even though you feel strong, take it easy.

Family Walks

Many parents who badly want their kids to love the wilds either cram it down their throats or fail to make those first trips really memorable. To ensure that your children enjoy a happy outing means tailoring it in all aspects to maximize their enjoyment. The same goes for other strangers to the wilds, whether girlfriends and boyfriends, children, parents, or even inexperienced friends.

Generally speaking, I feel parents should not take their children hiking until: (1) they themselves can travel in the wilds with some degree of comfort and competence; (2) the children want (or are at least willing) to go; and (3) the parents genuinely want them along. Families able to meet these criteria have a fighting chance for a pleasant trip.

The most common cause of disastrous family trips, it seems to me, is the failure of parents to see the trip through their children's eyes. Any child will ask, "If it isn't fun, why do it?" They do not insist every minute be fun, but they will expect that, taken as a whole, the trip should be pleasant.

After all, what good is a vacation if it isn't fun? The honest adult will find it hard to object to such logic.

Food that meets the adult backpacker's demands may or may not satisfy your children. Special emphasis should be put on snack foods and liquids. To maintain energy and prevent dehydration, children will need plenty of both. To encourage the between-meal eating so necessary in the wilds, plenty of goodies are needed, especially gorp, cheese, salami, nuts, and dried fruit. Children should be urged to drink small amounts of water often.

Parents unwilling or unable to keep the children amused and moving—without losing their own senses of humor—will wish they had left them home. So will the children! Without supervision, children tend to start out fast, which means they will soon want to rest. It is hard to curb the enthusiasm and effervescence that eats up their energy without curtailing the fun of the trip; it is also hard to keep them going when they feel tired and want to rest, and children fatigue quickly.

It is unfair to expect children to have the self-discipline necessary to conserve energy for the climb ahead. Instead, one has to supply incentives, distractions, goals, and just plain entertainment—with the minimum necessary discipline mixed in. Call the trip a "walk." "Walking" is fun but "hiking" is work. Making the trip a lark for the kids means getting into the spirit of *their* adventure rather than fretting about the slowness of the pace. Keep children moving but don't try to make them hurry; it will only slow them down and rob them of their cheerfulness. Make sure they get away from the trail occasionally. Let them go investigate something they've discovered.

Take them off the trail to see mossy glens, snowbanks, waterfalls, a tree that looks like a witch. If they've been trudging along wearily for a while, don't wait for them to ask for a rest or simply sit down. Stop voluntarily, give them something to eat, and show them something interesting. If they're happy they'll recover from their fatigue with amazing swiftness. Keep in mind the fact that the long-range goal is to make the trip so much fun that they'll want to come again.

It's always a good idea to keep watch on children's feet. By putting a stop to chafing in the early stages, you may avoid having to carry a child to the car. I have more than once discovered my daughter hiking happily along despite the fact that one sock had worked so far down her foot that it had disappeared entirely into her boot. Fortunately, children's feet take the abuse of rough country much better than their parents' and do not easily blister.

I find it important on the trail to talk to children a good part of the time. I give them progress reports: "We're more than halfway. . . . It's only fifteen minutes until lunch. . . . There's a spring where we can get a drink behind that big tree. . . . It's all downhill now." Whenever I can, I praise their achievements. I try to distract them from the drudgery of the trail and in doing so I find I have distracted myself.

When they grow weary of such temporal phenomena as birds' nests, rills, and rock rabbits, I try to stir their imaginations by pointing out a cloud formation that looks like a ship, a leaning tree that resembles a poised runner, or a patch of lichen that looks like a lion. Finding strange likenesses can be made into a contest in which children point out their own discoveries. The reward for the most imaginative can be a specially prized piece of candy.

I carry a considerable stock of snacks in a wide variety, and I keep them concealed to add mystery and anticipation. It's less important that kids eat well at mealtime than it is to feed them snacks between meals to keep their energy and spirits up. I pass out food with the smallest provocation and often with none at all.

Motivating Little Hikers

It's always important to keep the kids happy, but there's still the problem of getting them up the trail. A minimal amount of discipline, and self-discipline, is indispensable. So is desire on the children's part to please their parents. I explain at the outset that we're going to stop and rest, play games, explore, and generally have fun. As we move along, I show kids easier ways to get around obstacles, help foot draggers, readjust packs, show how the Indians walk, and if the trail grows steep, I demonstrate the rest step, which I represent variously as the "polar bear shuffle," "kangaroo limp," "dromedary drag," etc.

If there are several children, I work most with the slowest ones. Sometimes the slowest become the fastest if you put them in the lead, explaining that they now have the responsibility for keeping the group on the trail, showing them how to recognize blazes and ducks and the footworn groove. It is usually best to bring up the rear when hiking with children, so you can help the ones who fall behind and so you'll know if a child quietly sits down on a rock or wanders off while the rest of the party marches by.

The hardest parts of handling a group of kids are keeping them together and controlling the rest stops. Energetic older boys will want to keep going, while younger girls will frequently want to rest. It's not difficult to spread your party all over the mountain. I urge the stoppers to keep going and as a last resort I take their packs. If other kids get too far ahead, I may saddle them with the unwanted packs. When we stop for a rest I encourage the energetic ones to explore the immediate area while the tired ones sit and puff.

When the weariest seem to be somewhat restored, I simultaneously announce we must be off and pass out lemon drops. It is important that after a rest not to let children dash up the trail with recharged enthusiasm or they'll burn themselves out after only a few yards and plead for another rest. If they fail to restrain themselves, after you've explained the reason for starting off slowly, there is nothing to do but nag. Chronic fast starters are best reminded at the end of the rest instead of after they take off.

Even with all these stratagems, the trail can grow monotonous, and when the group becomes dull or dispirited I call an early rest. Everyone takes off their packs, and we make a little side trip to some interesting spot out of sight of the trail—usually a waterfall or a cool glade or a lookout point, and we have a drink of lemonade or lie in the cool grass, or throw snowballs off the cliff. This side trip is likely to refresh the group and the time spent seems a worthwhile investment. Progress, in this fashion, will be anywhere from a quarter mile to one mile per hour.

Limit the amount that thirsty children drink, but allow them—in fact urge them—to drink frequently from unpolluted streams and rills and your water bottle or canteen. Snowball fights and singing make good diversions. So do yodeling and echoing. Give kids the sense of helping you find the way. Don't communicate anxiety about snakes, storms, or mosquitos. You must be relaxed and at home in the woods if you want your children to feel the same way. Don't panic if they step near the edge of a cliff; their natural caution should protect them. Don't yell at them. Be alert for excessive fatigue, dizziness, blisters, chafing clothes, sunburn, and chapped lips.

Explain in advance that you want to keep them comfortable, and therefore they must let you know what's bothering them. They'll tell you all right! When you come to sand on smooth slab, wet slippery surfaces, loose gravel, mud, etc., calmly demonstrate how to cross safely. Celebrate all achievement; be liberal with praise and rewards. Teach your kids not to litter and get them to help you pick up gum wrappers and trash—if the going is not too difficult—to be deposited in one of your spare plastic bags.

Toilet Training

Impress your kids that nothing must be left behind to mar the wilderness, especially used toilet paper. Children at all timers should carry about three feet of TP folded up in a pocket. Teaching them when traveling to choose a place that won't be found; supervise if necessary. Show them how to fold the used paper inward for easy handling and put it in a plastic bag for later burning. Unless you provide instruction, there are sure to be toilet paper streamers decorating the trees, and the responsibility will be yours.

It's a poor idea to plan to take children cross-country in rough terrain unless they are large, strong, proven hikers. Being closer to the ground, children see relatively small objects as real obstacles. A rock that's just a knee-high step to you will be a waist-high roadblock to a six-year-old; a good scramble for you may be a nightmare for them. Forget boulder-hopping altogether.

Taking children into the wilderness can be demanding, even maddening, but by allowing yourself to see the trip through their eyes you can share their wonder, joy, and adventure. You can remember what it feels like to be a kid in the woods when everything is new and mysterious and exciting. And when the trip is over and you're homeward bound there is deep satisfaction in hearing your youngest ask, "When can we go again?"

Cut Your Toenails

If I'm getting ready for a serious dayhike or an extended trip, my preparations include cutting my toenails at least two days beforehand. Unfortunately, there's nothing harder to remember. My solution: I write it down on my To Do list for the appropriate day. If you've ever discovered on the eve of a trip that your toenails are too long, you know you're in trouble. If you cut them short at the last minute, you probably suffered for your neglect.

Freshly cut toenails can cut painfully into tender skin that's newly exposed. And failure to cut long toenails can bring another kind of pain when you descend a trail with what seem like short boots, each step jamming your toenails against the leather. So, somehow find a way to make sure that your toenails are appropriately short at least two days before you put on your boots for a real walk.

Get Off the Trail

A majority of hikers, or so it seems to me, are slaves to the trails. Many newcomers to walking are perhaps not aware that trails are the means, not the ends. The trail, however faint, is merely an extension of civilization. Wilderness does not begin until the trail is left behind. Many trip planners, without thinking, plot their routes exclusively from existing trail systems. And many squander a whole vacation on wilderness travel that never leaves the beaten path. But the solitude—the true wilderness experience—does not materialize until the traveler is finding his own way through wild country, rather than following a route marked by others.

By far the easiest way to escape the trail is to dayhike, carrying only sweater, lunch, and first aid kit. More country can be covered in a day without a pack than on a weekend under load. Hikers who want to spend a maximum amount of time in truly wild country may find it more fruitful to car camp close to the trailhead and spend the time dayhiking. I sometimes begin a wilderness weekend by backpacking cross-country for less than a mile to some unsuspected campsite by a spring or small creek; I then spend the bulk of my time dayhiking unencumbered.

A good many trips have been ruined in the planning by the seemingly harmless assumption "we ought to be able to make ten miles a day." On some days fifty feet would be too far. People have a habit of committing themselves to rigid goals: making 11.2 miles, fishing Lockjaw Lake, climbing Indian Peak. Somehow these achievements become substituted for the original or underlying reason for going—to enjoy roaming wild country. When people become so achievement-oriented that they measure the success of a trip in terms of miles tramped, elevation gained, or speed records, they often find themselves losing interest in wilderness travel. Working toward ambitious goals becomes too much like the rat-race at home. When peak bagging becomes obsessive or competitive, when each summit conquered becomes a mere achievement, the joy of reaching new heights disappears.

Trails provide a measure of dependability and security. Cross-country walking is altogether different. Instead of relying on an established course, one must find his own way; instead of the improved footing of a prepared trail, there are obstacles to contend with. Cross-country can be serious business and requires much greater experience, balance, strength, adventurousness, and caution than does trail walking.

In the space of a mile, one may have to contend with brush, bog, loose sand, boulder slopes, snow, deadfalls, mud, streams, and cliffs. And one of the most treacherous steep slopes I ever descended was covered with innocent-looking tufts of extremely slippery grass. Just as slippery are glacially polished slabs that are wet, mossy, or invisibly dusted with sand. Footing of this sort demands caution. I often take some trouble to climb around a wet or mossy slab, and when traction is vital I test the slope for sand by listening for the telltale grating sound. When I must cross slippery terrain, I often twist my foot slightly as I put my weight upon it to determine how well my boot soles are gripping.

When climbing a sandy slope it is important to plant the foot as flatly as possible; the greater the surface area of boot on sand the shorter the distance one is likely to slip backward. If there are rocks or patches of grass or low brush, I think of them as stepping stones and zigzag from one to the other. Sometimes steep sand is best treated like snow and the easiest way up is a series of switchbacking traverses or a herringbone step in which the toes are turned outward.

If the trail divides when we're exploring new country, my partner and I split up, each taking a branch, calling to maintain contact until we decide which one to follow. Often the split is only a brief detour and the trails soon rejoin. Other times we discover in a few hundred yards where the branch is headed, making a note to follow it another day.

Before starting out, be aware of likely hazards to be encountered en route—like poison oak and nettles, poisonous snakes, dope growers, mean dogs, angry property owners, livestock, hunters, dirt bikers, snowmobilers, fog, rednecks, private armies, shooting ranges, barbed wire, electric fences, swamps, rising tides, armed farmers—and take appropriate precautions.

Swing your arms when walking for added stability on treacherous terrain or when trying to make fast time. They'll help you balance as you dance past obstructions. When the trail takes a sudden dip, then climbs steeply, I often run through the dip so my momentum carries me up the far side, conserving energy and refreshing me with the use of a different set of muscles.

Beware hiking across slick surfaces with your hands jammed in your pockets. I was walking one day in snow that hid the ice just beneath, my cold hands jammed in the bottom of my trouser pockets. My feet went out from under me so fast that I hit the ground with my hands still stuck in my pockets. I landed on one arm, cracking three ribs, and was sore for a month. Now I automatically take my hands from

spring snow by a rock is often undermined.... so step over such areas.

my pockets, no matter how cold the day, if I suspect the slickness could cause a sudden fall.

If the trail turns steeply uphill or down or I'm picking my way across a sidehill, I stamp my feet very slightly with each step, especially in damp duff or loose rock, to gain extra traction by getting maximum shoe sole against the trail.

Hard and hummocky slopes of spring snow can be extremely tiring, and nothing short of wading tests the waterproofing of boots so severely. It is virtually impossible to keep feet dry. All one can do is carry several pairs of dry socks and keep dry footwear waiting at the car.

In the spring there is the constant danger of falling through a thin crust of snow with painful, even serious, results. There is hardly an easier way to bark shins, twist ankles, and even break legs. The margins of spring snowfields should always be treated with suspicion. So should snow-covered logs and snow from which issues the muffled sound of gurgling water. The best strategy I know for testing suspect snow is to kick it without actually committing any weight to it. If it withstands the kicking it can probably support my weight. Sometimes a big step or jump will avoid the necessity of stepping on what looks like rotten or undermined snow.

The Joys of Climbing

One of the greatest sources of joy I know is climbing to some attractive summit. It has been a long time since I called myself a mountain climber, but I enjoy getting to the top as much as ever. And I get just as much pleasure from walking up a little granite dome after dinner to watch the sun go down as I do spending all day working my way up a big mountain. The important thing to remember is that anyone can (and should) make his or her way to the top of an appropriate hill, ridge, or peak. I know of no better way to savor the wilderness.

Climbs are classed roughly as follows. Class I means following a trail or the easiest of terrain to the top, in any kind of footwear. Class II requires good boots and perhaps the use of hands on more difficult terrain. Class III requires route-finding skills, the use of hands, and the possible use of ropes to protect the climber during exposure to very steep slopes or cliffs. Class IV means continuous exposure requiring ropes and perhaps pitons to protect the climber. Class V and above is technical climbing demanding constant aid on the most difficult slopes.

Scrambling—Class II or III climbing that requires the use of hands, but not ropes—demands agility, good balance, endurance, and desire. Success may depend on the scrambler's ability to discover a feasible route by studying the slope during the approach and by consulting a large-scale topographic map. The basic rules for beginning climbers include: never climb alone; never go up a pitch you cannot get down; never climb on your knees; lean out from, not in toward the slope when exposure is great; and never take chances or attempt maneuvers that are beyond your skill.

Despite the need for caution, climbing can be enjoyed by most walkers, including children. Both my wife and daughter have climbed a number of peaks with me; my daughter made her first ascent when she was six. It is unfortunate that so many people think climbing means inching up sheer cliffs by means of ropes, pitons, and limitless will power. There is immense satisfaction to be gained in scrambling up peaks that demand little more than determination and offer no disconcerting exposure.

Climbing can be as safe as the climber cares to make it. As I come down a mountain late in the day, I remind myself that the majority of mountaineering accidents occur after three in the afternoon, and that twice as many falls happen

on the way down as on the way up. Expert climbers force themselves to descend with caution, thinking out difficult steps in advance to keep down the chance of injury.

Rock-hopping—crossing a boulder field by stepping or jumping from rock to rock—is probably the most demanding and dangerous way to travel in the mountains, but it is often unavoidable. I mentally try to keep a step ahead of my feet so when I run out of rocks I will be able to stop. I also treat every boulder, no matter how large, as though the addition of my weight will cause it to move. To slow myself down on a dangerous slope, I sometimes think back to a cross-country backpacking descent on which a companion, when forced to leap from a rolling boulder, opened six inches of his leg to the bone. Whenever I am forced to make a sudden or awkward jump, I try to land simultaneously on both feet with knees bent, to cushion the shock and minimize the danger of injury.

Baby Your Tootsies

Clean feet are happy feet. When you get dirt in your boots, your socks don't cushion as well and the dirt can cause chafing. Gravel can be infinitely worse. If a rock in your shoe is causing pain, stop immediately. Don't put it off. A rock will not only abrade the skin, it can make you favor the foot by walking unnaturally. And walking unnaturally can cause cramps, muscular aches, even injury. I generally pull the top of my socks down over my boots to keep out rocks, dirt, and sand.

And at every opportunity I take off my boots, turn my socks inside out to air in the sun and breeze, and wash my feet if I'm eating lunch by a lake or stream. I dry them with my bandana, letting the air complete the job. By the time I'm ready to go, my socks are much drier, my feet are dry and refreshed, and I'm ready to set forth in greater comfort—after emptying out my boots and knocking off the dust.

When I encounter loose sand, deep dust, fine gravel, or soft dirt, I remind myself to turn my toes inward. I know that my normal "toes out" style of walking will dependably shovel loose material into the back of my boots, where it trickles down to grate painfully on the back of my Achilles tendons. Consciously turning my toes in keeps my boots from shoveling, so I needn't stop as often to empty them out.

Nothing is more important to a dayhiker than his feet, the wheels that have to get him home. At the first hint of distress from somewhere inside your footwear, stop and attack the problem. Find the fold in your sock or the pebble in your instep. Loosen or tighten laces. You want your boots as tight as is comfortable. The tougher the terrain, the tighter they should be to give you maximum support. But it's better to have the boot or shoe top comfortable and loose than tight and chafing.

If there is soreness, take off both shoe and sock to find the problem. Inspect the skin carefully. The slightest pinkness demands attention. If you can't stop the chafing that's causing your foot to redden, put on adhesive tape, Moleskin or Molefoam, a bandage, or extra socks. Failure to deal with irritated skin will soon lead to blisters. Once blisters have formed the problem is quadrupled, because chances are good you can't totally prevent friction on the damaged area. And continued rubbing on blistered areas can mean real suffering. What if your feet hurt so much you can't put on your boots or can't walk back to the car? Prevention is the only solution to the problem of injured tissue on any part of the body.

Walker's Enemy: Chafing

Is there a fold in your shirt or sweater underneath the shoulder strap? Make sure you minimize chafing by paying close attention to all the points at which your body bears weight or there's friction of any kind. The lightest rubbing, if it continues mile after mile, can produce pain, irritation, blisters, or loss of skin that can become disabling. Maybe the irritation is slight at first so you procrastinate. But ask yourself: How will I hike tomorrow if I let a blister develop today?

Sometimes one shoulder strap is a tad tighter than the other. After an hour or so, this faint imbalance can result in an aching shoulder or neck or back. To insure maximum comfort at the end of the day (when you're sure to be tired anyway) be sensitive to the smallest discomfort. Then stop and do something about it *right now.* Veteran hikers baby themselves because they know what's coming if they don't. It's a sign of inexperience to hobble up the trail enduring discomfort. Concentrate on your problem and you'll discover your pack and clothing is more adjustable than you imagined. Experiment and you'll find ways to increase your walking comfort. And do it before your tender city skin turns pink.

Test Your Traction

Whenever I run into slippery footing, I want to know just how well my boots are handling it, to reduce the chance of falling. So whenever I find myself on wet smooth rock, slick mud, moss, wet wood, frozen mud, ice, damp clay,—anything potentially slippery—I pivot my foot slightly from side to side, while standing on the ball, noting the ease with which my boot slips. This traction test gives me an index of sorts to how cautiously I need to walk to safely avoid slipping—and perhaps falling.

After I've crossed a creek, waded through gumbo, sloshed through snow or ice, or skirted a mudhole, I know it's likely that the grooves between the treads of my boots are filled up, reducing my traction and increasing the weight of my boots. The best way I know to clear the grooves between the lugs is to sharply kick the nearest rock or tree. Clearing your cleats in this fashion will lighten your feet and restore the gripping power of your boots. Sometimes it's necessary to walk a ways on dry trail before a kick will dislodge the muck. And sometimes, of course, you have to sit down on a log and dig out the muck with a stick.

The "Limp" & "Indian" Steps

The single most valuable (and spectacular) walking technique I know of, one which literally flushes away fatigue, is variously called the "rest step" or "limp step." Though little known, this mountaineer's trick is based on the simplest of principles. When a hiker climbs steeply, the strain on the muscles around the knee is excessive and these muscles quickly fill with lactic and carbonic acids, the products of fatigue. This buildup of acids in overworked muscles, in turn, produces the painful ache that makes terrific slopes so uncomfortable.

The rest step is designed to flush away the acids of fatigue, thus relieving the ache they create. In the course of normal walking, knee muscles never quite relax. But if at some point in the step the leg is allowed to go entirely limp, even for only a fraction of a second, the excess acids are carried away and the pain miraculously disappears.

The necessary relaxation can be managed in either of two ways. The leading leg can be allowed to go limp for an instant just after the foot is placed for a new step and just before the weight is shifted to it. Or the trailing leg can be relaxed just after the weight is transferred to the lead leg and just before the trailing leg is lifted. I have gotten in the habit

of relaxing the lead leg, but most people seem to find it easier to let the trailing leg go limp. The trailing leg method is also easier to learn and teach.

My daughter learned it when she was eight. We were dayhiking up a relentlessly climbing trail that gains 1,200 feet in less than a mile. When she complained that her legs hurt, I had her stop and shift all her weight first to one leg, then the other, explaining that the pain would go away from a leg allowed to go limp. After she had stopped to flush her legs in this manner several times, I suggested that she take a small step forward with the leg that was relaxed, explaining that it was less tiring to keep going, even very slowly, when you rested your legs. Before we reached the top she was able to flush the fatigue from her legs whenever she needed to, without stopping.

As few as two or three limp steps in succession will bring amazing relief. Of course, the acids of fatigue continue to collect as long as the knees continue to work hard, so it soon becomes necessary to flush them again. But I find that after half a dozen limp steps I can return to my normal stride for anywhere from ten to a hundred yards. Besides offering relief from aching muscles, limp-stepping also provides comic relief by causing its practitioners to look like staggering drunks.

Another useful technique is the Indian step, a style of walking long used by cross-country skiers and European gymnasts as well as American Indians. Modern Americans tend to walk without swinging their hips. The Indian travels more efficiently. At the end of each step he swings the hip as well as the leg forward, pivoting at the waist. And he leans forward slightly as he walks. This forward lean and the turning of the hips lengthen the stride, position the feet almost directly in front of one another, and minimize the wasteful up-and-down movement. The result is a more fluid, floating walk, with less wasted motion. And on easy ground the longer stride produces more speed.

I occasionally use the Indian step if I wish to travel rapidly across level terrain that offers good footing. The easiest way to get the feel of the step is consciously to stretch the stride, thrusting the hip forward, aiming the foot for the center of the trail, swinging the shoulders counter to the hip thrust. Once the rhythm is established the shoulder swing can be reduced. Walking on narrow city curbs is a good way to practice.

Nothing consumes energy in such big gulps as maneuvers that require extra effort, like taking a giant step up onto a rock or log. If I cannot easily make my way around such obstacles, I transfer most of the extra effort to my shoulders and arms by placing both hands on top of the knee that is making the step and pushing down hard as I step upward.

On exceptionally steep rocky slopes, it sometimes becomes necessary to step forward onto the toe of the foot instead of the heel. Toe stepping adds power and balance on steep grades, but soon tires calf muscles. It helps to alternate heel and toe steps to prevent the cramping the latter produce. By following ten toe steps with twenty heel steps, I spread the work over two sets of muscles. The necessity of counting helps distract me from the rigors of the climb. If this arrangement continues to produce excessive fatigue, I sheer off from the fall line and climb in longer but easier switchbacking traverses.

When climbing cross-country it is sometimes necessary to remind oneself that the easiest route up may not be the easiest way down. Going up, I generally go out of my way to avoid sand, snow, and scree, but coming down I go out of my way to make use of them. Nothing is so pleasant after a hard climb up a mountain as glissading down a slanting snowfield or gliding with giant, sliding steps down slopes of sand or gravel.

Fording Streams

In the spring it can be dangerous to ford creeks and streams, never mind rivers. If you can't find a log or a series of stepping stones, you'll have to wade. The first decision is whether to protect your feet but soak your boots, or take a chance on injuring your feet (and increasing the likelihood of falling) by going barefoot. If you choose the latter, be sure your socks and boots are tied securely to your pack or around your neck as you wade, so that they can't possibly be lost if you fall or go under—unless you're prepared to walk barefoot to the car!

If conditions warrant, send the strongest person in the party across first with a rope. A fixed rope tied tightly between trees will provide security and peace of mind. So will a staff to probe for holes and brace like a third leg downstream against the current. Choose a wide shallow ford over a short but swift or deep one. The job will be easier if you start on the upstream side of a good fording site and plan to angle downstream, because that's where you'll end up anyway. If you can help it, don't cross immediately above a falls, cataract, or other substantial hazard. Organize your party for the safest possible crossing. And remember that risks are enormously magnified when you're traveling alone. It's better to change your plans or make any detour than it is to take a chance when there's no one to help if you get in trouble.

Use Those Arms

Most walkers don't use their arms enough. A vigorous arm swing functions like a gyroscope to help the walker maintain balance. The rougher the country the more valuable it is to swing those arms. If I'm boulder hopping or moving fast through rough country, my arms are going like crazy. I quite literally dance my way through the obstacles. And if I'm descending a steep loose slope through forest, I often swing from tree to tree like Tarzan. I safely descend a slope that would otherwise be dangerous by counting on catching branches or saplings to slow myself down.

Daydream Painkiller

Every experienced dayhiker at some time or other has experienced the sinking feeling of coming around a bend to discover a long, shadeless trail switchbacking endlessly upward toward a high and distant pass. When I find myself faced with a prospect of this sort, I often distract myself from the ordeal with the self-induced euphoria that comes of concentrated daydreaming. In a state of mild self-hypnosis, my day-

dreams so totally absorb my conscious mind that the discomfort or the grind goes mercifully dim.

As I start upward toward the pass I rummage about in my memory for some event or scene that is so thoroughly pleasant and engrossing that I recall it with consummate relish. Then I unhurriedly embellish my recollection with endless details that enable it vividly to fill my conscious mind. At first, it may be hard to escape into the past, but as the details pile up my awareness of present time and distance almost ceases. I climb automatically, sufficiently aware of my surroundings to make the necessary adjustments, but too engrossed with my dream to feel the discomfort. In fact, I'm sometimes reluctant, when the pass has been reached, to abandon my dream and shift my attention to the country ahead.

If I'm not alone I prefer the more companionable distraction of conversation. One of my regular walking companions, when we face a demanding stretch of trail, will say, "Well, what shall we talk about?" We may very well get rid of a quarter of a mile before we settle on a suitable topic. Often we trade accounts of movies, dreams, books, trout we have caught, or mountains we have climbed. Sometimes we may be driven to simple word games (especially useful with children) like Twenty Questions or Animal-Mineral-Vegetable. If we have been out in the country awhile, we may get rid of half an hour concocting menus for fantastic meals. Talking as we move upward tends to slow the pace, but that, in turn, further reduces the discomfort.

When I was a boy I had the good fortune to belong to the Boy Naturalists led by the well-known author-naturalist Vinson Brown. Vince used to take us into wild places he knew in the hills and station us, out of sight of one another, perhaps a hundred feet apart, on conveniently located rocks and logs. After wiggling into comfortable positions, we would be instructed to sit absolutely still for five minutes, not moving anything but our eyes.

If we were quiet enough, Vince told us, the birds and insects and small animals in the area would come gradually to accept us as part of the environment, just as they accepted the rocks or logs on which we were sitting. It was truly remarkable how well it worked. I don't know a better way to get close to the country, and often when I am walking alone in wild areas I will seat myself in some fruitful-looking place and let myself once again become part of nature.

Trail Manners

With more and more people walking in the wilds, trail manners have become more important. In most states discharging firearms, even during hunting season, is illegal across or in the vicinity of a trail. Equally objectionable is the boom of gunfire, which invades privacy and solitude and shatters the wilderness experience of other travelers for miles around. Guns are not needed as protection against wildlife and they have no place in today's crowded wildlands.

Horses and pack stock, once necessary to reach remote country, are now less common. But since stock can be unpredictable and difficult to control, it retains the right-of-way on trails. Walkers should move several yards off the trail, preferably downslope, and stand quietly while animals pass. Since walkers inevitably travel at different speeds, slower-moving parties should be considerate enough of faster walkers to let them move by. And fast hikers ought to politely ask permission to pass when the trail is narrow.

Overriding the hiker's concern for his comfort should be a sense of responsibility toward the country through which he passes. Increased travel in diminishing wild areas makes it necessary for all of us, consciously, to protect the environment and keep it clean. On the trail this means throwing away nothing, not even a cigarette butt, broken shoelace, or match. Never bury garbage. Leftover edibles, not including egg shells and orange peels, can be scattered for the birds and animals. Everything else should go in heavy plastic garbage bags to be packed out. The thoughtful walker takes pride in leaving no trace of his passing.

When it comes time to relieve myself I play a kind of game: I pretend to be a trapper who is traveling in the country of the sharp-eyed, sharp-nosed Sioux, and I hunt for a concealed spot unlikely to be visited by scouting parties—under a deadfall, in a thicket of brush, behind a boulder. I roll away a rock or kick a hole in loose soil, and slip the rubber band from the toilet paper roll around my wrist. When I have made my contribution, I burn the paper, cover the hole and jump on it once or twice to pack down the soil; then I roll back the rock, kick back a covering of leaves, duff, or pine needles and branches, artfully landscaping the site until an Indian scout would never give it a second look.

I am careful to avoid any place that might serve as a campsite or trail. Running water, dry streambeds, empty snowmelt pools, and any location within a hundred feet of a camp are also shunned.

Cleaning Up the Country

In recent years I have increasingly become interested in cleaning up the country. Restoring a quality of wildness to the wilds by erasing the blighting marks of humans has become one of my chief pleasures. It seems the least I can do to repay the joy that wilderness has given me and the one thing I can do to help make it last a little longer.

It also gives me a sense of accomplishment to restore an ugly site to its previous condition of natural beauty. For instance, a handsome little alpine lake may be marred by a sheet of glittering aluminum foil lying on the bottom. The act of fishing out the foil, crumpling it up and packing it into a corner of the knapsack restores the beauty and naturalness of the setting. Simple acts of this sort, surprisingly enough, can provide immense satisfaction.

I have found that people who endlessly procrastinate about cleaning out their cellars will conscientiously clean up other people's campsites. And children whose rooms at home stand knee-deep in litter will patiently pick the broken glass from someone else's fireplace. The fact that "trash begets trash" is as true in the wilds as it is in the city. Let litter accumulate and travelers will feel free to help swell the accumulation. But when a public place has been freshly cleaned, people become reluctant to scatter their trash.

I like to see how wild I can make a campsite look. There are far too many campsites already in existence; my aim is to decrease the number. After collecting all the unburned foil and metal from the ashes and packing it in a garbage bag, I bury charcoal, ashes, and partially burned twigs or scatter them well back in thick brush.

The firepit is filled in, blackened rocks are hidden and turned black side down in the brush, and bed and tent sites are filled in and smoothed over. Pine needles, sand, soil, duff, pine cones, and branches are scattered naturally about, and if the area still faintly resembles a camp I sometimes roll in a few rocks and large limbs or small logs to fill up the bare spots. After firewood is scattered, all that remains are a few footprints. Once rain has fallen it would be difficult to tell that the area ever had been disturbed.

cleaning up the lakeshore can be very satisfying

Safety Tips

Don't flirt with hypothermia. It's the number one outdoor killer. Take along enough clothing to keep warm under all conceivable conditions. I learned that the hard way on my closest brush with hypothermia. When I headed into Desolation Wilderness alone, it was an unusually warm summer day, without a cloud in the sky. So all I wore besides my boots was shorts and a thin T-shirt. But waiting just out of sight behind a peak was a storm. I didn't see it until I reached the high country.

Suddenly clouds blocked the sun. Then it started to rain, so I turned back. When the rain turned to snow, I started to run. Teeth chattering, soaked, losing body heat fast, shaking with cold and fatigue, I ran for miles. I didn't dare stop until I reached the warmth and shelter of a cabin. Even then it took hours by the fire to bring my thoroughly chilled body back to normal.

Safety also means guarding against lightning, falls, snakes, poison oak, biting and stinging insects, and so forth. Lightning kills nearly fifty hikers every year. On an easy climb of Silver Peak, I again was surprised by a hidden storm. Winds were high, but rain was light. Thunder was booming, but the lightning was striking more than a mile away. I felt I could safely make the last thousand yards to the summit, though my companion refused to come with me.

When I reached the top, lightning began to strike all around me. Boom and flash were simultaneous. I felt naked beneath a furious artillery barrage. There was no place to hide, but I didn't dare descend. I wedged myself half under a small boulder and waited, heart pounding, for the strike that would explode both me and the boulder. For twenty minutes I crouched there, soaked and shivering in the rain, until the bombardment shifted and I was able to run to where my partner was safely waiting.

I also learned the hard way why authorities admonish "don't climb alone." When I first went hiking above beautiful Zermatt, Switzerland, on my twenty-first birthday, I exuberantly began climbing an innocent-looking crag called the Gornergratt, planning to eat lunch at the restaurant I knew was on top. It seemed like the perfect dayhike. I didn't think of danger. It didn't occur to me that the restaurant might only be reachable by cable tramway. Climbing steadily, I expected to find a good trail at any moment, so I didn't pay close attention to my route. After scrambling up a twenty-foot cliff, I found my way conclusively blocked. But when I looked down, I found myself 2,000 feet directly above the Rhone Glacier!

If I fell while descending there was a fifty-fifty chance that I'd fall all the way to the glacier! Sweating, I started down, remembering too late that it's far harder and more dangerous to go down than come up. Halfway down the pitch I found myself blocked again. I couldn't find a needed foothold, hidden somewhere below me. There was nothing to do but jump. With all my strength I threw myself away from the abyss, clawing as I fell in the direction of safety. I landed two feet from the brink, bruised and scratched but intact. Since then, whenever I'm tempted to attempt a solo scramble, I remind myself of what might have happened the day I came of age on the Gornergratt.

Reading the Weather

Wind blowing from the south, east, or northeast is liable to bring a storm. Wind from the north, northwest, or southwest generally heralds fair weather; the major exceptions are mountain thunderstorms, which often come from the west.

Clouds offer another aid to weather forecasting. Big fluffy cumulus clouds are harbingers of fair weather so long as they do not join together and begin to billow upward. When they cease to exist as individual clouds, and the bottoms darken, and the tops form columns and flattened anvil heads a thunderstorm is on the way. High, thin cirrus clouds are generally filled with ice particles; when they whiten the sky or their mare's tails reach upward, a storm can generally be expected within twenty-four hours. Stratus clouds, as the name implies, come in waves or layers or bands. When they are smooth and regular and rolling the weather should be fair but probably cool; when stratus clouds are mottled or fragmented into a buttermilk sky, it will usually storm.

The astute wilderness traveler learns to recognize a number of signs of impending weather change. Sun dogs, or halos around the sun, forecast rain or snow; so does a ring around the moon. A red sky at dawn, an early morning rainbow, or the absence of dew on the grass—all of these should warn the traveler that bad weather is brewing. So should yellow sunsets and still, ominously quiet moist air.

Sensitivity to weather signs should influence the dayhiker's choice of camps and trip itinerary. If the signs are bad, he should erect the tent, lay in a good supply of firewood beneath a tarp, and schedule close-to-home amusement rather than exposed activities like climbing. On the other hand, a careful reading of the weather may enable him to set out on a climb, confident of good weather, even before the rain has stopped. For instance, clearing can be expected despite heavy clouds, providing "there's enough blue to make a Dutchman a pair of pants."

Storm Warnings

Thunderstorms, those exciting, dramatic, generally short-lived phenomena, are nevertheless frightening to a good many people. I have seen tall trees virtually explode when struck by a bolt of lightning, sending huge limbs flying in every direction. But the danger is negligible for anyone willing to take the necessary precaution—leaving vulnerable locations before the storm begins. The places to avoid are high, open, and exposed slopes, hills, ridges, and peaks; isolated or unusually tall trees, lakes, meadows, or open flats. The safest places are in caves, canyon bottoms, and a part of the forest where the trees are comparatively short.

The hiker or climber anxious above lightning usually has considerable warning. When cumulus clouds have darkened and fused, and still air has been replaced by sudden erratic winds, the storm is about to break and you should already be snugly sheltered. Lightning usually appears to be striking closer than it is, especially at night. The distance can be accurately gauged by counting the seconds between flash and boom. Every five seconds in time means a mile in distance. Thirteen seconds between flash and boom means the light-

ning is striking two and a half miles away. No matter how fierce the storm may seem, summer afternoon thunderstorms characteristically are short, and the chances are good that the sun will be out before sunset.

I always sniff around at the end of a dead-end road in good hiking country in hopes of finding the beginning of a trail. Half the time I'm lucky. To add some adventure to a walk in a tame city park, get off the trail and go cross-country for a wilder adventure. A small waxed paper cup can be folded flat to make an excellent drinking cup. Deanne and I regularly carry them in our daypacks.

The capacity of your daypack can be increased in dry weather by the addition of ties made of nylon line. Most often I use one as a clothesline for a freshly washed bandana or damp socks. I have used them to tie up trash I want to haul out or to lash on a garment that won't conveniently fit in my pack.

Large safety pins are equally useful. There's always one on my pack. Sometimes I tie (or pin) on a hat I don't need. String can be used to replace a broken shoelace. String is even more useful than a bandana. Don't gamble by simply tucking things in and hoping they stay, unless you're prepared to do without when you discover they fell off some miles back. And use square knots or slipknots, not bows, which untie, or grannies, which open up.

A dayhiker rarely needs to light a fire, but when he needs one he needs it badly. If you're caught by a rainsquall and drenched, you may desperately need a fire to dry out your clothes to stave off hypothermia. Kitchen matches are light and cheap and low on bulk. I've always got a few in the bottom of my daypack. Be sure you take the kind that can be struck on anything rough.

Finally, protect against canteen leakage by enclosing yours in a plastic bag (useable later for packing out trash). A friend of mine lost a camera when his canteen full of orange juice leaked on it all morning. And if you froze your canteen overnight, put it in a plastic bag before you put it in your pack. It will sweat a ton of moisture as it gradually thaws out on the trail.

TEN
Route Finding

It's Both Art & Science . . . Knowing Where You Are . . . Map Reading . . . The Compass Game . . . Telling Time by the Compass . . . Get Off the Trail . . . A Navigation Story . . . Finding the Trailhead . . . Gauging Your Speed . . . Getting Lost . . . Four Ways to Find North . . . Getting Found

Route finding is the art of getting where you want to go, whether you're in the middle of a city or on the flanks of a strange mountain in trackless wilderness. It means knowing where you are in the wilds, elegantly making your way through the country, evading obstacles and barriers. It's the key to successful cross-country travel, a substitute for blindly following the trail. Route-finding success will help you feel at home wherever you may be.

In the large sense, it's navigating through unfamiliar territory to a distant goal. On a smaller scale, it means figuring out how to get past a swamp or how to finesse a cliff with the least amount of time and effort. On an even smaller scale, it may mean choosing the easy way around a tree, picking the best rock to step on, ducking a branch, sidestepping thorns, deciding where next to plant your foot.

It's Both Art & Science

The exercise of route-finding ability, it will be seen, is both pleasant and practical. It can save your life. It can turn failure into success, convert a nightmare into rewarding adventure, prevent the anxiety, danger, and inconvenience of getting lost. It's the ultimate outdoor tool and challenge, problem solving at its finest, the zen of mastering the country.

For me, the most exciting, demanding, and rewarding aspect of traveling new country is pitting my ability to read the country (and use my map) against the dangers and enigmas of the natural world. Nothing is more deeply satisfying or more stimulating than successfully plotting a course through the wilds. I dearly love to sit on a ridge and plan my way through the country beyond, to scrutinize a map for clues to a cross-country route between two unknown watersheds, to examine the face of the mountain before me and figure out how best to get up it.

Sometimes route finding means finding the only way through a maze or around a cliff, solving the puzzle that makes the trip possible. Other times the benefits are more esthetic: looking for the prettiest, most appealing route or staying in the shade to hike in comfort. The satisfaction comes from the success, the dominion. Route finding isn't conquering the country so much as it is becoming one with it, solving its riddles, moving through it with ease, making it yours, going with the flow instead of against it, finding its hidden secrets and delights.

An old prospector with whom I used to travel in the deserts taught me never to "cut the country," that is, not to go against the grain. "Follow the canyons and the ridges," he advised. "The best footing usually makes for the easiest path, even if it is a mite longer." As we looked for gold, he showed me how to "go with" the country instead of fighting against it, to follow the ridges and the valleys and avoid "sidehilling." But valleys can turn into impassible canyons and ridges can narrow to knife-edge cliffs. Walkers must always try to conserve the elevation they have laboriously gained. A balance must be struck and that requires route-finding judgement.

The noted climber and explorer Jack Miller continued my education while we were traveling in the Andes of Bolivia and Patagonia. Studying the face of the mountain before us, he made me estimate distance, height, and elevation gain as well as projected routes and travel time. Then he'd correct my mistakes. Forced to make specific assessments and decisions, I gradually improved. Practice sharpened my judgement. Now I sometimes ask Deanne what route we should take on the slope that lies before us to make her focus on the problem instead of merely following my lead. Sometimes the route she picks is better than mine.

I feel richly content when I've cunningly made my way up a new mountain in interesting fashion—surmounting the obstacles, avoiding the pitfalls, discovering concealed treasures—and then found an even more rewarding route down. I've exercised my mind and used all my past experience to turn terra incognita into my own turf.

The pleasures of route finding are out there waiting to be discovered and enjoyed. I find them more seductive by far than mere egocentric peak bagging. As with most things, it's not just what you accomplish but how you go about it that brings the greatest contentment.

Knowing Where You Are

Route finding begins with knowing where you are. If you know where you are, by definition you can't be lost. The development of proficiency at route finding builds valuable confidence. It's the essential quality of outdoorsmen and women. The great explorers all have it. It will enable you to venture into strange country without getting lost. And if you become temporarily uncertain of your location, it keeps your from panicking and permits you soon to find your way again.

Even trail travelers need route-finding skills. It can be dangerous to put all your faith in the trail, blindly depending upon it to get you to your goal and home again. What if your trail inexplicably branches or disappears? What if it's wrong or a sign is missing, a junction unmarked? You're helpless if you've relied entirely on the trail. Even if you never plan to leave the well-worn, well-signed trail, you need to be able to figure out where you are.

Trail travel is fine if the path takes you exactly where you want to go, or if there's no other way through difficult country. Trails permit rapid travel, and because of the better footing they allow more attention to the scenery. But following trails can get dull. If you're a slave to trails you're going to miss the most stimulating part of traveling the natural world. The true adventure begins when you step off the trail. But you need to know where you are and be prepared to find your way.

Map Reading

The principal tool for knowing where you are is your map, though sometimes a guidebook or other written description of the terrain may adequately substitute. It seems obvious to say that a map is useless if you don't know how to use it. But a recent survey revealed that a majority of Americans today can't plot their way through a city with an ordinary street map!

Using a map means being able to read it, that is, recognize its features, understand its nomenclature, and be able to imagine and visualize what those features look like on the ground from the way they look on paper. It also means orienting the map so you're sure which way is up. If you can't line up your map with the country—or the city—it's useless. A map that's upside down is worse than useless because it gives false directions. Orienting a map means pointing its top toward true north, not magnetic north, or lining up known landmarks (like mountains) by sight.

The traveler simply spreads out the map with the compass on top so that north points directly to the top of the sheet. The map and compass are then turned as a unit until the north needle points directly to north. When the magnetic declination (difference between true and magnetic north) is set off (it is shown on all topo maps) the map will be oriented.

When one's position is known and the map is oriented, it becomes possible to identify visible features of the country-

she finds out where she is by triangulating from two known landmarks.

side by transferring line-of-sight bearings to the map by means of a straight edge. When only one's general location is known, but several landmarks have been positively identified, it is possible to discover one's precise location by transferring line-of-sight bearings to two known landmarks onto the map. The intersecting lines reveal the compassman's exact location. This won't work, of course, where convenient landmarks can't be seen.

Reading the rise and fall of the land on a topo map is more a matter of practice than talent. The best place to begin is with solitary peaks or hills where the contour lines form concentric circles that get smaller as they go higher toward the summit circle in the center. Widely spaced lines indicate a gentle slope, while lines bunched together describe a cliff. Contour lines form arrows that point upstream as they cross water courses and downhill as they descend a ridge or bluff. As one becomes adept at reading topo maps, the actual shapes of landforms begin to materialize so that the maps become pictures of the country.

When this happens, topo maps take on a singular fascination. They also become incredibly useful. For instance, by measuring the ups and downs of a trail one can estimate with fair accuracy the time that will be required to traverse a given section of country. This, in turn, allows the trip planner to work out logical camping spots and estimate the time needed for any itinerary. Superior map-reading ability is a prime prerequisite for safely visiting the wildest of trailless country. Proficiency with map and compass has saved a great many lives.

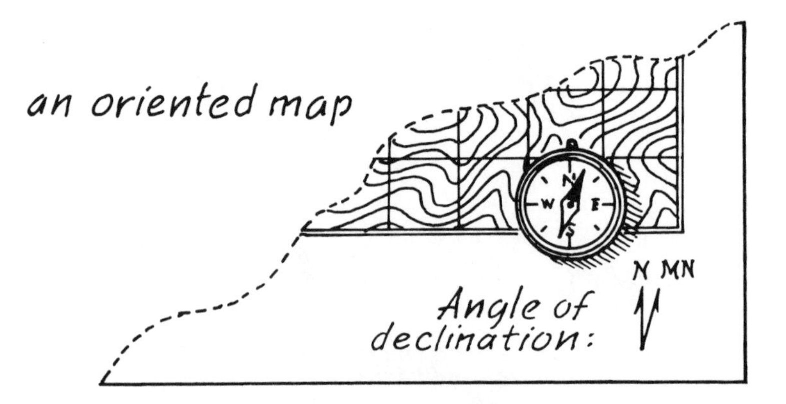

The Compass Game

The competent woodsman or mountaineer always tries to develop some inner orientation to make him independent of his compass. To this end, my friends and I play a game on the trail—while taking a rest—that has proven both amusing and instructive. Each of us draws a line in the dirt toward what he believes is true north. Then the compass is brought out, the declination marked off, and the winner declared. The next time the game is played the previous session's winner must draw first. It is surprising how quickly this game develops a sharp sense of direction.

Telling Time by Compass

If I really need to know the time, my compass will give a rough approximation, provided the sun is shining. I set the compass in the sun, settle the needle on north, then set off the declination (seventeen degrees east in California). With the compass thus oriented, I stand a straight twig on the compass rim so that its shadow falls across the needle hub to the opposite rim. The position of the shadow on the opposite rim gives me a close approximation of sun time by thinking of the compass as a watch with north at noon.

To reconcile sun time to daylight saving time, I add an hour. To tell time early or late in the day one only needs to know the hours of sunrise and sunset. Of course, some allowance must be made for mountains that rise high to either the east or west. Time is still important, but it is sun time, not clock time, that counts. How long before sunset, when it starts to grow cool? How long before dark? These are the pertinent questions in the wilds.

Get Off the Trail

Some hikers find a greater sense of adventure in deliberately planning no itinerary, going where and when the spirit moves them. Others, similarly motivated, spurn trails altogether in favor of cross-country travel, finding pleasure in avoiding people and the beaten path. Since true wilderness does not begin until the trail has been left behind, it should probably be the goal of most walkers to travel cross-country. Wilderness travel means getting off the trail.

I like to start thinking about future trips while I am still in the mountains. I find myself wondering, for instance, whether the rocky, trailless canyon I am passing could be ascended to the lake that lies above. As I move along, I try to estimate the difficulty of reaching passes, following ridges and crossing slopes, with an eye toward future trips. I study the topo map to learn what lies on the other side of the mountain and to compare the bunching of contour lines with those on slopes I can see or have already climbed.

This type of on-the-spot research becomes invaluable for plotting feasible cross-country routes in trailless country. A glance up a canyon shows the gentle slope suggested by the topo map turns out to be a series of ledges and cliffs. Or I confidently plan a route down a forbidding steep slope because I know it to be an easy sand and scree slide.

A Navigation Story

Careful trip planning, I've found, generally yields the most memorable trips and prevents the small disasters that can ruin a fine outing. Prospector Murl defined adventure as "the result of poor planning." Time has proved him correct. Successful trip planning requires route-finding skills. To plan a trip you must somehow picture it. After choosing your goal(s) you must estimate the distance to be walked and the elevation gain. To compute likely speed, you must consider the weather and terrain. Only then can you judge the time and energy required to make the trip.

These abstract considerations might be clarified by looking at a trip that three of us took while I was writing this chapter. It was the end of October in the California Sierra, and we hoped to squeeze in one last high trip before winter. We wanted to climb 10,400-foot Round Top Mountain just south of Carson Pass. None of us had been there but we'd seen the north side of the peak from a nearby summit, so we knew the type of terrain, the landmarks to the north, and the driving time required.

The fifteen-minute USGS topo map showed three trails that might be used to approach the mountain. Two of them started at opposite ends of Woods Lake at 8,200 feet, passed mountain lakes, and joined near the mountain's west ridge. There the contour lines were well spaced, suggesting a feasible route up. We decided to ascend by one trail and return by the other, making a loop. The trail distance looked to be about six miles, with a climb of 1,200 feet. The ascent of Round Top would require another mile each way, cross country, with another 1,000-foot elevation gain.

We judged we could cover the eight miles, with a total 4,400-foot elevation change, fairly comfortably in five hours. We wouldn't have to leave early and we'd be home in time for planned Saturday night socializing.

The weather forecast warned that a storm was moving in from Alaska on Sunday. Snowfall could be heavy and close the area for the winter, so we set forth on Saturday morning, leaving at a civilized 9:00 A.M. Although we planned to hike in shorts, we took long pants, watch caps, and mittens in the car, just in case. And I brought my vapor barrier shirt. Less than an hour's drive from our home in Pollock Pines, we reached Woods Lake.

We were mildly surprised to find that a dusting of snow from a storm two weeks before still covered the shady north side of the peak, down to about 9,000 feet. But the sky was clear and sunny and the temperatures comparatively mild, so we didn't hesitate. We had come prepared. The mittens, long pants, and watch caps went into our daypacks. Since the creeks would be flowing with snowmelt, we left all but a pint of water in the car. We set forth up the trail about 11:00 A.M.

When we emerged from the trees at timberline and started dodging snowbanks, it was windy—advance disturbance from the approaching storm—but the air remained mild and we walked in shorts. We made fast time on a good trail, passing choppy Winnemucca Lake and climbing to where the trail crosses the mountain's north ridge at 9,400 feet. The flat sunny ridge offered a largely snow-free corridor that led us to the high west ridge, where fifty-mile-an-hour winds filled our eyes with tears, spun us around like staggering drunks, and made us put on mittens, watch caps, and wind shells—but not long pants.

Gaining the sunny south slope behind the ridge, we left the snow behind and made our way to the windy summit, taking shelter wherever we could. After climbing all morning on fruit power alone, we ate lunch in a crow's nest behind sunny sheltered rocks just under the crest while the wind roared inches above our heads. Then we swiftly descended on loose scree to the trail, drank snowmelt from a stream and took the homeward trail, via Round Top Lake, back to our car at Woods Lake. We were back home by 5:00 P.M. in plenty of time for the Saturday night festivities. Careful planning and route finding had made the trip a success. The following day it snowed.

Finding the Trailhead

Sometimes the toughest part of route finding is finding the trailhead. I've wasted a lot of time trying to get off a freeway, driving up and down dirt roads, thrashing through underbrush—all in search of some elusive trail. Sometimes I merely hoped to find a trail, even a game trail, heading in the direction I wanted to go. At other times, the directions I'd obtained were bad or vague. I've learned that I've got to allow time to find a strange trail. If I can, in new country I scout the trailhead beforehand—the day or night before—so I don't waste precious time on the day of my hike.

Gauging Your Speed

The question inevitably asked by strangers to the wilderness is how fast will I travel? Or, how far should I plan to go in a day? There are no answers, only generalities. On relatively level trails at moderate elevations, a long-legged, well-conditioned man may manage four miles per hour. At elevations over 6,000 feet in rolling country, a well-acclimated walker is moving extremely well if he can cover three miles per hour. The average hiker, fresh from the city, heading up into the mountains will be lucky to average two miles per hour. These speeds are for people who keep moving and should probably be cut in half for those who want to poke along, smell the flowers, take pictures, and enjoy the view.

Cross-country walkers may average only one mile per hour. Children, the elderly, and hikers strongly affected by the altitude may manage as little as half a mile an hour. A friend of mine has devised a useful formula for predicting his speed. He plans every hour to cover two miles if the trail is flat or mildly descending. For each thousand feet of rise he adds another hour. For steeply descending trail he adds a half hour per thousand feet, increasing that to a full hour for extremely steep downhill trails. By plotting a rough profile of the trail from the topo map and applying his formula, he can estimate quite accurately the time needed to walk it.

There is no particular virtue in covering great distances. I have received more pleasure, solitude, and sense of wilderness from hiking less than two miles cross-country into a neglected corner than I have from following sixty miles of well-traveled trail.

Survival should not be a problem for the dayhiker unless he is hurt or lost. A first aid kit and the ability to use it should enable him to cope with all but the worst accidents and health failures. His experience and trip preparation should enable him to survive whatever bad weather or minor mishaps befall him with no more than discomfort. It simply isn't possible to protect against all the hazards of wilderness travel. Develop self-reliance, build your experience, and don't make the same mistake twice.

Getting Lost

Getting lost—or thoroughly confused—is not uncommon in the wilds and rarely leads to tragedy if sensible procedures are followed. I have been unsure of my location a good many times without suffering unduly. The lost hiker's ability to regain his sense of direction and rediscover his location (or make himself easy to find) depends largely on his ability to control panic and fear so that logic and reason can prevail.

The best insurance I know against getting lost in strange country is to study it as I move along. I consult the map and reorient myself with each new turning of the trail. I identify and study the configuration of new landmarks as they appear. I frequently look backward over the country just traversed to see how it looks going the other way. If I am traveling cross-country or on an unmapped trail, I stop several times each mile to draw the route on my map. And at critical points (stream crossings, trail branchings, confusing turns), I make appropriate entries in my notebook or on my map.

A compass is the easiest means of orientation, but the sun, the stars, and the time can also be used to determine direction and to act as a check on the compass. More than a few disasters result when lost travelers refuse to accept what their compasses tell them. Unaware that they are lost and convinced they have their bearings, they assume the compass is broken or being unnaturally influenced.

Incidentally, precision compasses rarely are needed in wild country simply because it's impossible for the hiker to measure distance accurately and maintain precise bearings. One degree of error in a mile amounts to only ninety-two feet, so compasses that can be read to the nearest five degrees will be accurate enough for most purposes. In strange country where landmarks are unfamiliar, a compass is no less than indispensable. Many times I have worked out my true location or avoided a wrong turn by referring to a dollar-sized compass that weighs a fraction of an ounce and cost less than a dollar. But don't forget to allow for the declination. A compass will be useful in well-known terrain if sudden fog cuts visibility. So will an altimeter.

Four Ways to Find North

A roughly accurate watch will function as a compass on any day that the sun is out. If one turns the watch so that the hour hand points toward the sun, true south will lie halfway between the hour hand and the number twelve. An allowance must be made if the watch is set on daylight saving time, which is generally an hour earlier than standard (or sun) time.

If the time is known generally, even if within only two hours, a rough but very useful idea of direction can be obtained simply by knowing that (in western America) the summer sun rises a little north of due east, stands due south at noon (standard time), and sets a little north of due west. In the winter the sun's path lies considerably to the south, rising south of east and setting south of west. Early and late in the day, we can easily determine directions with sufficient accuracy for most purposes.

Another useful strategy involves pushing a stick vertically in the ground, marking the end of its shadow, waiting about fifteen minutes, then marking the shadow again. A line between the two marks will run east-west (the first mark is the west end). A line drawn at right angles will run north-south.

If the night is reasonably clear (in the northern hemisphere), it is relatively easy to find the North Star (Polaris), which is never more than one degree from true north. Its location is determined from the Big Dipper (or Big Bear), a bright and easily identifiable constellation nearly always visible in the northern sky. A line drawn upward from the outermost stars at the bottom and lip of the cup will point to Polaris. In the southern hemisphere, the long axis of the Southern Cross points toward a starless region that lies due south. The prominent constellation Orion lies in a nearly north-south plane, and the uppermost of the three stars in the belt rises due east and sets due west from any point on the face of the earth.

Getting Found

There are so many variables in every situation that it is difficult to advise the walker who has managed to get lost. However, a few general rules nearly always apply. The novice hiker has a tendency to plunge on through country that has gradually grown unfamiliar in hopes of reaching a familiar landmark. The veteran will resist this impulse, stop, admit to himself that he is at least temporarily lost, and sit down to review the situation. When he has overcome the anxiety that often accompanies such an admission, he will rationally review the situation, carefully considering all the information available.

After studying the map and thinking carefully, he may find a landmark he can identify that will reveal his approximate position. Or thinking back over the country he has traversed he may feel that by retracing his steps he can return to a known point in a comparatively short time. After all the evidence has been sifted, the important decision is whether to try and return to a known point, whether to stay put and await rescue, or whether to head hopefully toward civilization. Only full consideration of the situation in a rational, panic-free manner will reveal the best course of action.

But prevention, of course, always beats cure. To avoid getting lost, hone your route-finding skills. Learn to read and orient maps. Play the compass game until you can always point to north. Practice navigating your way through the unfamiliar areas of country you know close to home. Develop an awareness of distance by making an association between map miles and hiking miles. Test your estimates of altitude and elevation gain against your map. Plot routes through and around local obstacles. Study topo maps until you can "see" the country they depict at a glance, and memorize your declination. As your route-finding proficiency begins to grow, you'll enjoy the deep satisfaction that comes with feeling at home in the wilds.

Adventure Guides, Inc., 36 East 57th St., New York, NY 10022. (212) 355-6334.

Alpine Adventure Trails Tours, 783 Cliffside Dr., Akron, OH 44313-5609. (216) 867-3771. Swiss Alpine rambles.

Alpine Expeditions, Box 1751, Bishop, CA 93514. (619) 873-5617. Camp-to-camp dayhikes.

Andean Outfitters, P.O. Box 220, Ridgeway, CO 81431. (303) 626-5918. Premier guide Jack Miller.

Country Walking Holidays, 1122 Fir Ave., Blaine, WA 98230. (604) 921-8304.

Fit For Life and *Fit For Life II: Living Health,* by Harvey and Marilyn Diamond (Warner Books, New York, 1987).

Hahnemann Medical Clinic, 1918 Bonita Ave., Berkeley, CA 94704. (415) 849-1925. Fine classical homeopathic care.

Hahnemann Pharmacy, 1918 Bonita Ave., Berkeley, CA 94704. (415) 548-5015. Source of safe, potent homeopathic remedies.

Hike Maui, P.O. Box 330969, Kahului, HI 96733. (808) 879-5270.

Homeopathy, Medicine That Works! by Robert S. Wood. Send $9.95 plus $2 shipping to Publisher Services, P.O. Box 2510, Novato, CA 94948. CA residents add 62 cents sales tax.

Knapsack Tours (Kathy Harrison), 5961 Zinn Dr., Oakland, CA 94611-5003. (415) 339-0160. Inexpensive group dayhikes.

Mountain Travel, Inc., 1398 Solano Ave., Albany, CA 94706. (800) 227-2384. Worldwide adventures, giant catalog.

Natahala Outdoor Center, U.S. 19W Box 41, Bryson City, NC 28713. (704) 488-2175. Walking tours in the South.

New Zealand Travelers, Inc., P.O. Box 605, Shelburne, VT 05482. Dayhike with guide Alan Riegleman.

Nichols Expeditions, 590 North 500 West, Moab, UT 84532. (800) 635-1792. Friendly outfitters Chuck and Judy Nichols.

1991 Adventure Travel North America, by Pat Dickerman (Henry Holt & Co. $15.95). Hundreds of outdoor outfitters.

Palisade School of Mountaineering, P.O. Box 694, Bishop, CA 93514. (619) 873-5037. Skywalks.

Sierra Club, 730 Polk St., San Francisco, CA 94109. (415) 776-2211. Thousands of outings each year.

Skoki Lodge, P.O. Box 5, Lake Louise, Alta, Canada TOL 1EO. (403) 522-3555. Base camp for walking trips.

Sobek Expeditions, P.O. Box 1089, Angels Camp, CA 95222. (800) 777-7939. Worldwide treks, big catalog.

Ten Speed Press, P.O. Box 7123, Berkeley, CA 94707. (415) 845-8414. Publisher of *Mountain Cabin, Whitewater Boatman, Pleasure Packing, 2 Oz. Backpacker.*

Vermont Hiking Holidays, P.O. Box 750, Bristol, VT 05443. (802) 453-4816. Hike inn to inn.

Walking Magazine, 711 Boylston St., Boston, MA 02116. $12/yr. Bi-monthly. Publishers of an annual *Walking Source Book* for $2.95.

Walking—The Pleasure Exercise, by Mort Malkin (Rodale Press, Emmaus, PA. 1986). A 60-day fitness program.

Wilderness Travel, 801 Allston Way, Berkeley, CA 94710. (800) 247-6700. Worldwide adventures, giant catalog.

Yosemite Mountaineering School, Yosemite, CA 93514. (209) 372-1335. Guided hikes.

Yosemite Park & Curry Co., 5410 East Home Ave., Fresno, CA 93727. (209) 454-2002. Yosemite backcountry high camps.

About the Author

Long a seeker after carefree adventure in the wilds, Robert S. Wood is the author of the backpacking best-sellers *Pleasure Packing* and *The 2 Oz. Backpacker.* Wishing in recent years to walk every day—instead of just a few times overnight in the summer—he turned his attention to dayhiking.

He has been hiking for forty years in his beloved Sierra, the Cascades, Mexico, Europe, Alaska, New Zealand, South America, and Australia. As a journalist he wrote and edited for a number of magazines, including *Life, Time, Sports Illustrated, Sierra, Wilderness Camping,* and *Outside.*

Since fortunate investments permitted early retirement, he has devoted his time to wilderness travel, here and abroad, river rafting, hiking, and personal growth—and writing eight books about his adventures. A native of Berkeley and a forestry graduate of U.C., he now divides his year—with wife Deanne and daughter Angela—between homes in the Sierra foothills and the Big Island of Hawaii and a summer cabin on the edge of Desolation Wilderness.